SHEPHERD

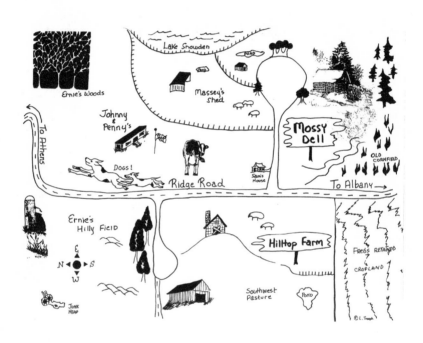

SHEPHERD

A Memoir

Richard Gilbert

Michigan State University Press
East Lansing

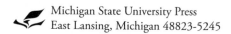 Michigan State University Press
East Lansing, Michigan 48823-5245

Printed and bound in the United States of America.

20 19 18 17 16 15 14   1 2 3 4 5 6 7 8 9 10

LIBRARY OF CONGRESS CATALOGING-IN-PUBLICATION DATA

Gilbert, Richard, 1955–
    Shepherd : a memoir / Richard Gilbert.
        p. cm.
    ISBN 978-1-61186-117-4 (pbk. : alk. paper)—ISBN 978-1-60917-407-1 (ebook) 1.
Gilbert, Richard, 1955– 2. Sheep ranchers—Ohio—Biography. 3. Sheep farming—Ohio.
4. Family farms—Ohio. I. Title.

    SF375.32.G54A3 2014
    636.3'1092—dc23   [B]   2013025307

Book design by Scribe Inc. (www.scribenet.com)
Cover design by Shaun Allshouse, www.shaunallshouse.com
Cover image is of Freckles and two of her lambs, and is used courtesy of Richard Gilbert.

green press INITIATIVE Michigan State University Press is a member of the Green Press Initiative and
is committed to developing and encouraging ecologically responsible publishing
practices. For more information about the Green Press Initiative and the use of recycled
paper in book publishing, please visit www.greenpressinitiative.org.

Visit Michigan State University Press at www.msupress.org

For Kathy, Claire, and Tom
And to the memory of my parents:
Charles Churchill Gilbert
Rozelle Rounsaville Gilbert

What tickles the corn to laugh out loud, and by what
star to steer the plough, and how to train the vine to elms,
good management of flocks and herds, the expertise bees need
to thrive—my lord, Maecenas, such are the makings of the song
I take upon myself to sing.

—from *Georgics*, Virgil (Peter Fallon trans.)

SHEPHERD

PROLOGUE

CHILDHOOD DREAMS CAST LONG SHADOWS INTO A LIFE. AS IF THE strong feelings they stir prove their validity, dreams propel the dreamer through an indifferent world. Which explains how I, a guy who grew up in a Florida beach town, find myself crouched beside a suffering sheep in an Appalachian pasture.

"Richard, I think you should call the vet," says my wife. Kathy and I flank the ewe's prostrate body.

Our third lambing has just begun this spring of 2001, and Red is in trouble. I'd found the little ewe in distress and had urged her up and nudged her inside an old shed, where she'd collapsed and resumed straining, panting as if in labor. But nothing happens; no lambs, hour after hour.

Kathy knows I'm reluctant to seek paid help. We're on a tight budget and I'm trying to be a practical farmer, even if still part-time: commercial shepherds do their own veterinary work, or they simply shoot and compost ailing ewes. Profit margins, razor thin, can't support farm calls.

But Red, small and fine-boned, white with a roan patch on her neck, is a special case. She emerged four years ago from the anonymity of our new flock, fifteen rambunctious ewe lambs, by insisting on making herself our pet, surprising us and astonishing her wild flockmates. During a disastrous renovation of our farmhouse that summer—we maxed out our credit cards, spent our kids' college savings, and borrowed against our retirement

1

accounts—she'd sidle up every afternoon to be petted. As our other sheep stared at her in wide-eyed horror, Red mooned up at us with trusting eyes, charming us and lightening our cares.

So I relent and call Maggie Swenson this Sunday afternoon. She arrives as heavy shadows from hickory and locust trees fall across the pasture. Maggie announces after a quick check that Red's cervix isn't dilated, and then she sits back with a puzzled frown on her elfin face. Red sure appears to be in labor. I mention ringwomb, when a pregnant ewe comes to full term yet fails to dilate for delivery, which I've heard about in an e-mail listserv for shepherds.

"Could be," Maggie says doubtfully, her short gray hair luminous in the shed's deepening gloom. I wonder if ringwomb is even a recognized sheep ailment—there are so many, and I can't find it in my reference books—or just more gossip from my virtual colleagues, who keep me fretting about the endless woes that sheep are heir to.

"Should we load her up and bring her in?" I ask.

"I don't see what good it'll do to bring her in."

Maggie seems to be writing Red off, though she's too kind to say anything so harsh, and her attitude puzzles me as much as Red's condition does. Where's the help I'm paying for? "If she hasn't delivered by tomorrow," I say, "maybe I should bring her in for a C-section."

Maggie nods. "Call first," she says. Then she shows us how to massage Red's cervix, which Kathy does for an hour without result, and we return to our house and our children in the mild April darkness.

My father lost our farm in Georgia when I was six, and I grew up in Florida fantasizing about reclaiming our land. Decades later, when Kathy's job took us here, to a place where we could buy a working farm, my boyhood dream resurfaced and carried me away.

I'd said I wanted an adventure when we moved to this beautiful backwater, Appalachian Ohio, from our comfy house in prosperous suburban Indiana, yet the first thing I learned was that I didn't like starting over in a new place; it was *hard*, and didn't suit my need for stability. As a local acquaintance jokes about such matters, "It weren't easy." In early middle age, Kathy and I had returned to Ohio with two young children—not to the state's busy capital where we'd met in graduate school fifteen years before, but to Athens, a battered brick enclave in the state's remote southeastern corner. We were shocked. Athens seemed to have more in common with a village in dusty western Kansas than it did with long-domesticated Ohio, a state of orderly farms, dotted with big industrial cities. Just before moving I had learned that Ohio had an Appalachian region, and upon arrival saw what that meant: poor, tired towns; roadsides choked with litter; the land abused for decades by extractive industries—coal mining, timber cutting, natural gas drilling; the growing of erosive row crops like corn. I was horrified that we'd traded our stable, affluent world for this pinched place.

We'd moved only six hours east from Indiana, but had come so far. We couldn't know how far we had to go. We hadn't expected to find such stark regional differences. We'd accepted the myth that America has been homogenized, scrubbed clean of warty local distinctions by affluence, by shared television shows, and by broadcasters whose bland voices erode the patterns of proudly regional speech. Part of me was thrilled to discover there are still *places* in America, and for all its poverty and isolation it was the most beautiful landscape I'd ever seen. In Ohio's hill country, a wrinkled shirttail dangling untucked above West Virginia, everything felt different: the layered woods, the light flashing off pebbled creeks, the wind in the trees, the wild phlox that bloomed pink beside shaded roadsides late in May.

All we needed to make my dream come true, I'd vowed to myself after our exhausting relocation, was our own land. I pictured cattle lowing across a drowsy green valley in the sun-heavy afternoon, a red banty hen clucking in the barnyard dust. What I hadn't known as a daydreaming kid in Florida, and couldn't foresee as a neophyte agrarian in Athens buying land and livestock, was that my Eden could be so very complicated. All I knew, though I wasn't sure why, was that I had to act on my desire at last.

On Monday, as I drive across the road to where Red is confined, I hope to find her nursing twins, or maybe triplets. I'll turn her out into the bright sunshine with her lambs and she'll graze the shiny spring grass.

Stranger things have happened. In our first lambing, everything went wrong: ewes rejected lambs, two ewes died, and we had to pull a lamb that had expired in the womb. Last year there had been only one big problem, but a doozie, unheard of in a mature sheep. During the strain of late pregnancy, a ewe suffered a rectal prolapse, meaning her rectum fell out; it hung down like the trunk of a baby elephant and I had to shove it back in—four times. I concluded that severe genetic faults afflicted most of our ewes. Now Red is down. For the hundredth time I curse her lackadaisical breeder. And guiltily remember my excitement when I bought our first ewe lambs at such a bargain price, only $100 each.

I park and march uphill to the shed. Red lies where Kathy and I left her last night, and I kneel on the hay beside her. Both of her sides bulge with the splayed saddlebags appearance of a ewe hugely swollen in late pregnancy. Her eyes tightly closed, she saws her outstretched neck from side to side as if trying to get the right angle, to find some comfort. She grinds her teeth in pain. She moans, a human sound, and moves her head upward, lost inside her struggle.

"Oh, girl," I say. Poor Red. I can't maintain a farmer's resignation—I'll have to get Maggie to try surgery.

As I return home to give her a call, I remember what Kathy said as I left our farmhouse this morning: "If you decide to take Red to the vet, don't try to load her by yourself. Ask Sam to help you." Our neighbor Sam is always eager to help—true. A retired handyman for the university where Kathy and I work, he takes an avid interest in my activities. But I don't want to get started with him, not yet, not so early in the farming year. I need to lamb alone. The ewes know me, I tell myself, and I don't want them spooked by a stranger—anyone they don't see daily. In truth, I can't face goodhearted Sam's constant questions, which would be as maddening as gnats around my eyes during this latest, mystifying crisis.

As I drive my pickup truck back up our driveway, I feel again like such an amateur. I've set so much in motion that's beyond my control. Yet when things go right, I know, lambing season is almost unbearably exciting. My pocket lambing notebooks, cheap spiral-bound Oxford booklets from Wal-Mart's shelves, capture, as if enchanted, its chaos, drama, and abundant living gifts. Even now, years later, I can turn their dog-eared pages—smudged by dew and birth fluids, smeared with blood and dirt—and hear lambs crying and their mothers baaing and see a ewe licking her newborns in the sun. The chunky notebooks contain data: birth dates, weights, tag numbers. And scrawled exhortations: *Great mother!* or *Cull this ewe!* My records affirm that Red was a calm, patient mother in her first two lambings. She never fled from me, dragging her panicked newborns behind, or lost them, upsetting other ewes by baaing her way through the flock looking for them.

I park at our house, call the veterinary office, and try to figure out how to load Red. Although her breeding weight is only about 115 pounds, she probably weighs an additional twenty-five at full term. As svelte as a gazelle, she doesn't *look* 140 pounds, even bloated by pregnancy. Small-boned myself, I weigh about 165. Although my legs are strong, my upper body isn't, and at age forty-six, there's no doubt I've inherited Dad's bad back, which leaves me sore and creaky after a hard weekend of farm projects and sometimes lays me low with painful spasms.

I know to be careful, to lift with my legs. I'll get Red to town with a short lift into our big two-wheeled garden cart and then another into the bed of my truck. As I roll the cart to my pickup, I suppress an image of Red panicking and thrashing in my arms, her flinty hooves spearing my stomach as her iron-hard head clobbers my face. Our fifty new ewes regard me warily from the pasture beside our house. In March I bought them in a flock dispersal, a great chance to expand with a strong new bloodline, but now I'm feeling overwhelmed, with new chores and new worries about having enough grass and hay to feed two flocks totaling one hundred ewes. Since getting the new sheep, I've found myself chewing my fingernails as I drive to my day job each morning.

I've taken two weeks of vacation for lambing, starting today, and have plenty of time for tending ewes, delivering lambs, and handling the odd emergency. Yet my vague but persistent fears visit me, magnifying Red's emergency. My busyness and the greedy project I've hurled myself into have devoured my spare time. I feel guilty for neglecting Kathy and our children, and I know my coworkers think I'm crazy. And what do I really know about farming, anyway? Practically nothing. After Dad sold our Georgia farm,

I grew up a block from the Atlantic Ocean; the only sheep for a hundred miles was pictured in grainy black and white in one of Dad's ancient textbooks. As I heave the cart aboard my truck, anxiety spikes through me.

A decade later, a friend will ask me about our Appalachian adventure: "Was your experience typical? As a beginning farmer? All the problems?" I'll wonder what she really wants to know. About our house disaster? The puzzling birthing problems in our first flock? The vile disease that surfaced in our second? Being caught between feuding neighbors? My injuries? My emotions? *Me?*

With a guilty pang, I pass Sam's tidy house for the third time this morning, the wheels of my upended cart spinning slowly in the bed of my truck. I wish I didn't have to deal with this crisis, although by now I know that something is always going wrong on a livestock farm.

But God, now *Red*. This feels so cruelly wrong, so upsetting. So lonely.

"Ask Sam to help you," Kathy had said.

PART ONE

LAND TIES

It takes a long time and a great deal of effort to understand the quirks, peculiarities, and nuances of another culture. How much truer when we are dealing with an entirely different species than ourselves!

—Burt Smith, *Moving 'Em:
A Guide to Low Stress Animal Handling*

CHAPTER ONE

Walking up and down these hills is an education in itself.

—e-mail to a friend

KATHY HAD FOUND THE FARM YESTERDAY, IN THE GENTLE SNOWFALL of our first Appalachian winter. Now she drove me to her discovery. Before we were out of town, I peppered her with questions and steeled myself for disappointment.

It was December 1996, five months since we'd moved from Bloomington, Indiana, to Athens, Ohio. Five hard months of searching for a farm—either the houses or the land, or both, had been wrecks. One of Kathy's secretaries, who'd grown up outside Athens, had told her that a farm would be auctioned to settle an estate. Immediately Kathy had been hopeful: a place that the woman casually mentioned, as if everyone knew it, really was going to be sold, without realtors or owners involved.

"How much land again?" I asked.

"Seventeen acres."

Just twice the size of the farmette we'd owned in Indiana, far smaller than I'd hoped for. Another remnant, when I wanted a full square. I frowned.

"Only *seventeen*?"

"You've got to see it, Richard. It's magical."

"Maybe there's adjoining land." But I knew the place wasn't going to be right.

Kathy fell silent. We took the arrow-straight Appalachian Highway, a robust four-lane that lay almost deserted between Cincinnati and Athens, to Albany, a crossroads west of Athens. Seven miles and we turned right onto a spur that delivered us into Albany's faded downtown—a video-rental and tanning shop, a funeral home, some vacant red-brick storefronts, a plywood pizza shack, a Hocking Valley Bank branch—and we followed a rural byway out of town. In only half a mile, in a sharp curve, Kathy angled our van onto Ridge Road, where the landscape opened up. Yesterday's low gray ceiling had lifted, and the cloudless sky, almost purple, was incandescent above the season's unsullied first snowfall.

Kathy turned onto Snowden Road, a narrow lane that threaded past corn stubble on one side and modest homesteads on the other; we passed through a woodlot and into a gravel farmyard that ended at a small pond. She swung right to park in front of a gnarled maple at the base of a hill. Above, a log cabin built of black logs, with wide white calking between, faced us. A high canopy of branches overhung the turnaround, and hills rose on each side. Cradled by the hills, beneath its bower of trees, the farmstead slumbered.

I jumped out into the thin snow and was startled by the slam of my door in the stillness. I took in the placid pond and the mossy cabin, and I gaped at the old trees. This was a place out of time. The flat gravel farmyard was a courtyard with many entrances: the road, the farm's lanes, and the doors of outbuildings that perched on the surrounding slopes. The buildings were chalky white, with streaks of dove gray that matched their pewter tin roofs, and constructed from stubbled planks.

We climbed two flights of stairs to the cabin. Its front door, painted dark brown, was made from impossibly broad boards, over two feet across— braggin' boards, a carpenter I'd met would say. We tracked through the powdery snow around the structure, noting the hundred-foot tulip poplars spared for shade behind, on the south side.

Back at the landing, we paused. The maple tree where our van was parked had entwined its branches with those of a massive three-legged white oak at the edge of the pond, and the limbs of the trees arched over the clearing like an umbrella. At the entrance to the farmyard, forty-foot hemlocks sent their feathery, downward-sweeping evergreen branches to meet the oak's and maple's. Across from us on the facing bank, an allée of enormous oaks—another ashy-trunked white oak and two with black boles—stood before a barn with a tall gambrel roof that brooded over the ice-rimmed pond.

I looked at Kathy. "That white oak on the bank must be two hundred years old," I said, "and those hemlocks are *huge* for such slow-growing trees. I half expect to see elves and fairies dancing under them. And a druid worshipping under the oak."

"I told you!"

I shrugged and grinned. We skipped down the stairs and crossed the farmyard to a one-vehicle track that angled toward a rusty woven-wire fence; beyond, a pasture rolled uphill. I pushed open a gate and we followed a tractor's wheel ruts. Halfway up, we turned and looked back. The rear of the barn was sunk into the slope so that hay wagons could be unloaded into its commodious roof. Across the farmyard, another lane climbed between the three-legged oak and the grandfather maple, past an eighty-foot white pine in the cabin's yard, and curved behind a low stable and disappeared into another high pasture.

"I think I see how the land breaks out," I said. "There are about five acres in woods at the farm's entrance. The farmstead's probably two. There are about five acres in this field, and there must be another five over there, behind the cabin. That's seventeen."

"Isn't it lovely?"

I drew a breath of cold air and looked at her—broad smile, brown eyes bright, and cheeks rosy from the cold; her thick brown hair, shiny and soft, touched the shoulders of her red sweater. "Yes," I breathed. "*Yes*."

At the top of the hill we saw a cylinder of greenish hay, as big as a hot tub, beached near the open mouth of an unpainted loafing shed; beneath its sagging tin roof, the shed's manure-fouled depths were as dark as a cave. The hay had been chomped and the ground around it churned to black mud. We glimpsed cattle eyeing us from downslope, near an overgrown boundary fence. Below the line fence, water flashed—the shimmering blue reach of a lake.

"Snowden?" I asked. During our farm search we'd frequently circled Albany's Lake Snowden, a reservoir that supplied a rural water company; living in town, on city services, we received regular warnings to boil our water because of line breaks. And to think this farm bordered such beauty, 140 acres of sparkling water.

Kathy grinned wider. "Surprise."

I called the lawyer listed on the ad we'd clipped from the Sunday *Athens Messenger* and asked him about the upcoming auction of the little farm Kathy had found.

"No, it's not a public sale, per se," Ted Foote said. "Sealed bid. Offers mailed to my office are due March 31, postmarked that day."

"My wife and I drove out yesterday. It's nice."

"Gorgeous," he said in a voice that was oddly both abrupt and affable. "That was Kenneth and Mabel Vaught's place. Kenneth ran the feed store in Albany. He's been dead ten years, and Mabel died last spring. Lost Valley was their retreat. Put their heart and soul into that place.

"We've already got a pile of bids here," he went on. "Everybody around Albany knew and loved Lost Valley. Of course, with the lake right there, that's rare. A farm with waterfront."

"Well, thanks, Mister Foote. We'll be bidding."

At lunchtime I threw on my coat and raced across campus to Kathy's office. I filled her in as she zapped leftovers for us in a microwave. "We need some help," she said. "That's not typical Athens County farmland."

We both knew whom she meant.

So that afternoon I took a deep breath and called our realtor, Brian Winesap, and told him about the sale. Brian had already shown us dozens of properties we'd rejected.

"I've heard about that place," he said. "I'll take a look."

"What will we owe you?"

"I've been meaning to get out there anyway."

When I was sixteen, browsing in a mall bookstore in Florida, I discovered paperback copies of Ohio writer Louis Bromfield's *Pleasant Valley* and

Malabar Farm. They are two of the most romantic books ever written about American agriculture, but I didn't know that then. I saw their color covers—images of dewy green pastures and freshly turned black loam—and grabbed them off the rack beside the cash register.

I was growing up in a tropical paradise but grieving for our Georgia farm. Dad hadn't been able to earn enough from farming to support his growing family, and in 1961 he'd gone to work at the Kennedy Space Center, during America's race-to-the-moon buildup. He still received the *Progressive Farmer*, and I'd grown up paging through it. In a way that dry magazine never could, Bromfield's books showed me how I might redeem the loss of our farm. A similar exile had shaped Bromfield's boyhood, but he'd returned as an adult to re-create the realm where he'd spent idyllic days. *Pleasant Valley* opens with a scene of him driving into a snowy valley and imagining the dreamy summer landscape he'd known as a child:

> What I saw was a spring stream in summer, flowing through pastures of bluegrass and white clover and bordered by willows. On a hot day you could strip off your clothes and slip into one of those deep holes and lie there in the cool water among the bluegills and crawfish, letting the cool water pour over you while the minnows nibbled at your toes. And when you climbed out to dry in the hot sun and dress yourself, you trampled on mint and its cool fragrance scented all the warm air about you.

Sometimes, curled up with the chunky mass-market paperbacks in an overstuffed chair in our Florida room in Satellite Beach, I ached with my sense of loss. When I excitedly showed Dad my books, surprise flickered across his handsome, impassive face—The Great Stone Face, a brother would one day call it—and he pointed out the hardcover originals in his library. Bound in black cloth, they were embossed with a red Harper & Brothers logo showing a torch being passed from one hand to another.

Later, as we sat in front of the TV, I raved about Bromfield's innovative sustainable practices. Dad was, as always, as concise and unsentimental as a telegram: "He didn't have to make money from farming."

Dad knew that Bromfield had supported Malabar Farm with money he made writing movie scripts in Hollywood. I knew from Mom that Dad had plowed his own fortune, an inheritance that might have made us wealthy if invested wisely, into two cattle ranches, the first in California after the war. He followed experts' cutting-edge advice, bought new tools to carry it out, and went his own solitary way. He ignored my mother's alarm over expenses and over some of his industrial-farming practices, such as injecting steers with growth hormones, that she sensed were dangerous.

What spoke to me as a broody kid in our Space Coast suburb was Bromfield's passionate quest, the emotional satisfaction of reclaiming land and restoring it to beauty and abundance. A man could create his own paradise, safe from crass commerce, if only he worked with nature. I dreamed away sunny days, lost in his humus-fueled vision, making it my own. Ever since I was six, when we'd left Stage Road Ranch in Georgia, I'd wanted a farm, and Bromfield gave me a vision of it: a place shaggily beautiful and fertile and sheltered and safe.

By the time I got to Dad's copy of Bromfield's final farming book—*From My Experience*, published in 1955, the year I was born—the exotic world of his Malabar Farm was as real to me as our beach town. Reading in our home a block from the Atlantic Ocean, I learned of the existence of a true paradise: Ohio.

I majored in journalism at college, like Louis Bromfield, but minored in agriculture at my father's suggestion—he thought I might write about it—and took a part-time job on the University of Florida's pig farm. After graduation and four hectic years reporting for newspapers in Georgia and Florida, I won a Kiplinger journalism fellowship to Ohio State University, in the center of Ohio, in Columbus, just south of Bromfield's legendary farm. I was hungry to pause and read books, but part of my impetus for taking a sabbatical year from the *Orlando Sentinel*, where I was covering the Kennedy Space Center, sprang from my fading agrarian dream. I yearned to see the fecund wonder that was Ohio and to visit Bromfield's farm, now a state park. As soon as I'd unpacked my bags I drove to Malabar Farm and was overwhelmed on a tour by the scale of the landscape, even though I knew it was a fraction of what Bromfield had farmed. With other tourists I drifted through the Big House and heard the guide's spiel about the day Humphrey Bogart and Lauren Bacall exchanged vows in its parlor.

Over the years I'd discovered other agrarian writers (Wendell Berry, Eliot Coleman, Gene Logsdon), and others lay in front of me (Wes Jackson, Joel Salatin, Allan Savory), but none could replace Bromfield as my emotional touchstone.

In Columbus I lived in a sooty brick row house; up the street was a biker hangout and a place to sell blood plasma. Before classes started I wandered Ohio State's vast campus, across sunny greens and beneath pools of shade from exotic northern trees. I wondered what winter would be like—I'd never been outside the South. I was twenty-six, my reddish-brown hair beginning to thin. I brought with me a gift from my mother, a six-month-old black Labrador retriever, Tess. Every evening I hurled a Frisbee into the Olentangy River, which passed through campus, and Tess flung herself into the water after it.

At a faculty reception that fall, I noticed a tall woman with a friendly Midwestern face, lustrous dark brown hair, and warm brown eyes. Later I made excuses to visit her office, and as we talked, tiny blushes warmed her cheeks. She radiated goodness, that breathtaking mystery. Kathy Krendl was thirty and had grown up on a farm in northwestern Ohio. After four years as a high school English teacher in Ohio and another four in graduate studies at the University of Michigan, she was lecturing at Ohio State while she wrote her dissertation. Later I learned she was feeling alone in the world, her parents having died young—her mother when she was a junior in college, and her father within the last year.

At the end of my fellowship year, we began dating. We visited Malabar Farm, and my dog charged into a swamp after a muskrat. Kathy and I held hands, laughed, and watched Tess go.

Tess's intensity sometimes alarmed Kathy. "She looks like she wants to eat me," she said.

I had a ready answer for once: "No, that's love."

I followed Kathy to Bloomington, Indiana. We arrived with nothing, but Kathy worked harder than anyone I'd ever known, harder, even, than my workaholic father. And she'd urge me out of bed earlier than I wanted, having gotten home from the copy desk of the *Indianapolis Star* at two in the morning, so I could apply my new Ohio State master's degree, teaching journalism part-time at Indiana University.

We had Claire in 1986; Tom arrived in 1988. We moved into our gleaming white colonial-style house, built to our specifications, in March 1990, just before my thirty-fifth birthday, a scant seven years since our arrival in Bloomington.

Surprising Kathy—and myself—with my release of a pent-up energy, around our faux farmhouse I planted hundreds of trees, shrubs, and perennial flowers. I grew a big vegetable garden and kept a flock of hens, Ameraucanas, whose pretty blue and green eggs I delivered to Bloomington's health-food store. I began to breed my own strain, saving eggs from the best layers to incubate in our basement. The city built a new elementary school right down our road, where Claire and Tom started school.

But after six years in our dream house, Kathy became restless. As the only dean with offices on each of Indiana University's eight campuses, she was caught amid constant faculty power struggles, and she'd stopped getting backup because the president who'd promoted her had left the university. When I told my mother that Kathy was looking for a new job, she went silent on the phone. Maybe I wanted her to argue with us, to say we were crazy to leave our prosperous world, but she'd never do that. Dad had dragged her from California to Georgia to Florida chasing his dreams, and she herself was flexible, able to make a home anywhere.

Honestly I was surprised by Kathy too, by her determination to start over. But, all the same, I found myself dreaming of a place with more land. Soon she signed on as dean of what Ohio University touted as its "jewel in the crown," its College of Communication. And I accepted a job with the university press, which needed a book publicist. By then I was working at Indiana University Press, one of the Midwest's largest academic publishers, and I reasoned that at a smaller press in Athens I'd have time to try farming on a larger scale.

Having put herself through college by growing and selling vegetables on her family's farm, Kathy had no interest in returning to commercial farming herself—on any scale. Hard labor had eclipsed any romantic notions about agriculture. Her father, the son of poor immigrant farmers, had worked relentlessly, always holding down two jobs in addition to farming.

Yet given her father's example, my yearning wasn't truly odd to Kathy; anyway, another country place would be great for the kids. And she responded to urgency and desire. "I don't have dreams like you do," she said to me one evening as we sat in our breakfast nook, taking a break from packing for our upcoming move. "I envy people who do. I just take things one opportunity at a time."

So we'd made our deal: in exchange for our uprooting, a big new job opportunity for Kathy, who now officially became the family's main bread-winner; a real farm for me; and an adventure for all of us. Our friends were dumbfounded that we'd leave all we'd built. We already had it made.

And soon my concerns about our destination grew. Using the Internet to research southeastern Ohio, I learned that over half the area's popula-tion lived below the poverty line; Lyndon Johnson had unveiled his Great Society program there in 1964. Although a regional oasis like Bloomington, Athens was smaller, with only 21,000 permanent residents; income from about 20,000 transient students drove the local economy.

We hadn't been able to find a farm during a quick family trip to Athens. So while I'd stayed with the kids one June weekend, Kathy had driven back to Athens and bought us a house in town to live in while we searched for a farm. The temporary place was secluded, Kathy assured me: a rustic retreat on five wooded acres. Balancing demands, as always, she'd passed up nicer houses for one that would accommodate our menagerie, which included our collie, Doty, Claire's four cats, and chickens.

I couldn't part with all my layers. That August I moved chickens to Appalachia.

One evening, still awaiting our realtor's report on what to bid on Lost Val-ley, Kathy heard a knock on our front door. I'd taken the kids to a karate class. She found on our doorstep a short man, with fiery red hair and a belly as taut as a beach ball, grinning through a scrim of red beard.

"They call me Fat Man," he said, gripping a clipboard. "You wanted an estimate?"

Yes, Kathy remembered—about lighting. She led him inside. The elderly couple we'd purchased our new house from had relied on a few dim 40-watt bulbs; they'd swum through an appalling gloom. Even now, lighted by Kathy's collection of ornate antique lamps, the interior remained dark. Overhanging trees shaded the windows all day, and the sagging porch that wrapped two sides blocked more sunshine. The funky house had been a Girl Scouts headquarters—hence the colors: the roof brown, the clapboards canary yellow—and the boarded-up hatch in the living-room wall was once the campers' take-out food window. Athens had since circled the retreat, a forgotten enclave astride a forested rise at the end of a suburban lane.

"I saw your picture in the paper," Fat Man said. "Do you want to know the word on the street about you?"

Alarmed, Kathy looked at him. Newspaper coverage, as if to stoke regional resentment, had emphasized the cost of an influx of new administrators. "No," she said. "No, I don't want to know."

As Fat Man beamed, our rooster crowed, sounding crazily loud in town, even in the woods at the end of a cul-de-sac.

"You'll never believe what you just missed," Kathy said, greeting me as I came in, climbing the stairs from the house's lower entrance. Claire and Tom plopped onto the couch in the basement family room to watch television. The kids had had a bruising experience because I was trying to replace their Hoosier karate school with its Appalachian equivalent. They'd expected to find a mentor just like theirs in Bloomington, a jokey "clown prince of karate" sensei with his wife assistant, a preschool teacher. But the Athens instructors had called them sissies and ordered them to hit and kick other students and to receive blows in return; I'd been running errands and hadn't witnessed the sparring.

"Let's give it a chance, kids," I'd muttered, driving.

But Claire, her brow moist, her brown eyes staring defiantly at the windshield of my truck, had blurted, "Those people are jerks! I'm not going back." Tom, our watchful child, his pale cheeks flushed, his blond hair stuck to his forehead, listened.

Now Kathy recounted her meeting. "Ignore him," I said. "We have enough to worry about without small-town gossip." I yanked open the refrigerator and found it almost empty.

She turned to me. "Did you hear there's another warning from the city about water?"

"*What?*"

"I heard it on the radio," she said. "Cooking and drinking water should be sterilized by boiling."

"That's the third order this week to boil the water."

"Second."

"We never had *one* in Indiana. What's wrong with this place?"

"The pipes are old and break," she said. "I guess they don't have enough money to fix them."

"I'm getting a bumper sticker made: Athens: Third World Living at Its Finest."

"I'm sorry," Kathy said. She paused and added, "That's not funny, Richard."

Our Indiana phase, thirteen lucky years, had gone so fast. And we'd been happy, it was so obvious now. Then, in an instant, Kathy's ambition and my dream had united. The sign had gone up: For Sale.

That night I telephoned my friend David Bailey. He'd been a Latin and Greek scholar who'd left academe for the grubbier but more exciting life of a journalist. We'd worked together at a newspaper in Cocoa, Florida, and he'd since returned home to North Carolina, where he wrote for a magazine.

"Bailey, you won't believe the newspapers here," I said, and held forth on their superficiality.

"*G-i-l-b-e-r-t,*" he drawled, "it hurts to be a fat cat, doesn't it? I feel your pain."

"What are you talking about?"

"I've seen you skewer people. Now you've got a big bull's-eye on *your* ass." He was laughing at me—actually chuckling.

"Remember," he went on, "when you wrote about that geezer who ran for state representative? You tore him a new one."

"He deserved it! He fell asleep at campaign rallies!"

"*You* deserve it," he brayed, "as far as anyone there knows."

No doubt I'd annoyed him with my giddy e-mails, pre-move, about our relocation to a rural paradise where we could live on a farm and have good jobs in town. "I can tell you've already drunk too much bourbon to absorb reason, too," I said. "But speaking of noxious liquids, we have to *boil* our water or bacteria will kill us."

"*G-i-l-b-e-r-t.* You don't sound like a *farmer*. Look what you've done to yourself. You had a cute place but wanted a *farm*. Welcome to the real world. Suck it up, yuppie."

I put down the phone and found myself chewing my fingernails. I thought of Dad. Since our move, I'd been marveling at his ability to move through unfamiliar worlds, driven by his dreams, consumed by his plans, conquering disparate realms. He'd shown up in sleepy Leesburg, Georgia, in 1957 with a wife and three young children. And I knew that place well enough to know how odd he'd been there.

When we lived in Florida, Mom would take us kids in the summer to visit our old Georgia hometown. Playing with children there, I knew that Leesburg was where I should've been growing up instead of Satellite Beach. "Listen," a boy said to me one night, "that's hounds on a trail down on the Muckalee Creek." My sense of having been torn from that dreamlike world of woods, fields, and coonhounds baying in the swamps almost brought me to tears. By the time I was a teenager, the differences between my Georgia peers and me—their lovely syrupy accents, their laughter over their wild partying, their bemused ease around grownups—had become too stark for me to enjoy visiting at all.

I'd imprinted on the landscape, however, on the vast fields relieved by dark islands of pines and bordered by creeks. My earliest memories clustered there: Dad brushing silently past me into the white farmhouse as I stood looking up at him; Mom showing me a long rattlesnake someone had lynched from an oak's high limb for all to see; dust rising behind a red tractor as Dad tilled. And my terror when Mom left me and my sister, Meg, in the car and descended the banks of the Muckalee to pick magnolia blossoms; that's where—*has she forgotten?*—she once showed me the path an alligator scuffed as it dragged one of Dad's newborn calves into the black water.

As a college student, feeling uncomfortably rootless, I'd driven to Leesburg myself. We'd been gone fifteen years by then. In the town's only grocery, I told the proprietor that we'd lived on the Stage Road Ranch. The man thought for a moment. Then he said that my father's ability to estimate the weight of a pen of market steers was the most remarkable skill he'd ever seen. I was proud that Dad's competence as a cattleman was still honored. I suspected that he was remembered less fondly there by my mother's close circle of friends.

Afterward I'd sworn never to return, because of the pain of loss it stirred up, but I visited again in middle age when a car trip took me through southwestern Georgia. I'd begun my nascent farming apprenticeship in Indiana by then. I'd begun to wonder about my father and his clashing dreams of flight and rootedness. In Leesburg I talked about him with a local couple as we ate fried quail, biscuits, and grits at their hunting lodge on property that bordered our old farm. The woman was the daughter of the mechanic who'd serviced Dad's tractors. The man was heir to this land, a peanut and cotton plantation whose most lucrative enterprise now was leasing quail-hunting rights.

"My Daddy said he'd never seen a farmer bring his tractors into town for service," the woman said, looking accusatory.

"Mine told me he'd never seen a white man who worked like your father," her husband said, sounding aggrieved.

They stared at me across the table, waiting. They were demanding, in an implicit southern way, that I explain him. Who knows what they'd actually heard from their parents—they were my age—but undoubtedly it boiled down to Dad being a crazy Yankee. He certainly hadn't socialized in the local way, drinking, smoking, and telling funny stories about his and other's foolish behavior. He hadn't tried to ape a southern accent or change his behavior to fit in. He must have seemed aloof and humorless.

In Satellite Beach I'd watched him charm outsiders, clients he had to entertain for work, or distant relatives, unaware of his nature, who were passing through. His blue eyes sparkled with humor; a smile lit his face. The current of warmth that flowed from him at such times was palpable, the way the Gulf Stream off our beach coursed in a warm vein through the Atlantic's murky coastal chop. Always he commanded respect, a natural leader, and when he dropped his sober mask he could be as charismatic as a movie star. His act angered me—a betrayal of my real father, the one living with the unfathomable loss of our farm—yet it also seemed a genuine expression of joy he'd buried within.

He never spoke of his deepest loss, the tragedy I'd unknowingly inherited, the suicide of his own father when he was fourteen.

Mom, who'd grown up poor in Oklahoma, had been accepted in Leesburg. She'd made a good impression right away by dressing up us kids every Sunday and marching us into the Baptist church. That led to her participation in dove suppers and bridge games. These were clannish people, but Mom was fun. She liked to laugh, was good at cards, a great cook, and wore nice clothes. She enjoyed *them* and their stories. No, she wasn't the question on the table.

I searched for words to help this couple place my father, words that would define his deeply private nature without implying that it had anything to do with them, because of course it didn't. I felt the usual pang of guilt for discussing him at all, never mind critically, with anyone outside the family. He needed interpreting, but doing so felt like a betrayal. Explaining my father without diminishing him would take hours, days, a lifetime.

At my Athens barbershop on Wednesday, still impatiently awaiting word from our realtor, I looked out at the lunchtime uptick in traffic as Ernie Tyler worked on my hair, talking nonstop. Beside us, his son, Jim, a man about my age, lounged in his own barber chair, smoking a Marlboro and squinting at a copy of *Western Horseman*.

Although my bald man's fringe didn't give him much to work with, Ernie was talented. Today, for some reason, he was cutting my hair with an old-fashioned straight razor, which, showcasing his casual skill, lent subtle drama to his work and a flourish to his gestures. My mind had wandered, and I wasn't following his story about a professor he'd beaten up. Apparently the fisticuffs were over politics, but seemed to have more to do with Ernie feeling disrespected by the town's ruling elite.

"Jim was cutting his hair and I was sitting where you are," Ernie said, moving in front of me, fixing me with his large black eyes. He held his silver razor delicately in his right hand, like a wand. "I told him he was full of shit. He got up and said something and threw his glasses on the counter like this"—Ernie mimed the man's contemptuous toss—"and I came out of the chair and knocked him down right where I'm standing. I held him with my left arm and punched him in the jaw with my right. I was trying to break his jaw but couldn't, just smashed in his teeth."

"You got the guy's attention!" I said, marveling at Ernie's encounters, and vowing to keep my political opinions to myself. His stories always seemed at least partly a test of me, an outsider who'd blundered onto his turf. Maybe he sensed my discomfort with his violent tales, because during my visits he'd gentle like a racehorse stabled with a harmless donkey.

"Speaking of lawyers," I said, seeing my opening, "Kathy and I are thinking of bidding on a farm in Albany. A lawyer here in town is handling the estate—"

"Lost Valley!" Jim exclaimed and dropped his magazine into his lap.

"It's right down the road from me," Ernie said, straightening my head and looking into the mirror over my bare scalp.

"I practically grew up there," Jim murmured. "In summer, a bunch of us boys would ride our ponies to Lost Valley every morning and chase each other up and down the road."

I pictured a younger Jim riding with abandon, bareback, pounding down Snowden Road, racing other boys, a band of whooping savages in the rising dust.

"When I was older," he said, "I rode hundreds of miles with Mabel. All the trails and back roads. In the afternoon I'd ride into Albany to Kenneth's feed store. Everyone said he got Lost Valley from a farmer who couldn't pay his feed bill. Kenneth had one of those old coolers in his store where the drinks were in water, ice-cold, and I'd get a pop and we'd talk horses. He ran a boarding stable right there in town. He used to haul the manure all the way out there and spread it on that farm."

"I can see why they called it Lost Valley," I said. "There's something about that farmyard. It's just a gravel turnaround, but with the trees and hills all around, you feel like you're in another world."

"That's not why," Jim said, crossing his legs, right cowboy boot atop left knee. "Before they built Lake Snowden, there was a valley on that farm. Margaret Creek ran through it. Oh, it was beautiful. So green and peaceful. That driveway in front of the barn used to be a little township road that crossed the bottom. People would take Sunday drives, look at the hayfields and the crops."

"Goddamn government bureaucrats," Ernie said. "Claimed they had to impound the creek for flood control. Paid everyone peanuts. They drowned our best corn ground. It's at the bottom of Lake Snowden. I had a nice field in that valley, and now all I've got on that side of Ridge Road is a strip of woods and a couple steep hills."

"The lake cut Lost Valley in two," Jim said. "The Vaughts sold off the other side because they had to drive around the lake to get to it. After that sale and what the lake took, they'd lost most of their land. Still, it was such a fun place. Kenneth and Mabel had horse shows and bonfires out there. I can still see them playing cards at the table in the cabin."

"What happened to their daughter? The lawyer told me she died."

"Yeah, she was just a kid," Jim said. "She got hit in the stomach with a softball. I guess she was complaining that night, so they took her to the hospital. But she died."

Jim, taller than his feisty father, with sandy hair and a contemplative mien, fell silent and stared out the shop's dusty plate-glass window. Ernie & Jim's Barbershop felt faded, timeless, a remnant from another era. For thirty years it had overlooked this downtown street corner. A radio muttered on the counter: on Sale-a-Thon, someone was trying to sell Chow-mix puppies.

Ernie had told me his life story on my first visit, as soon as I'd mentioned we were newcomers wanting to buy a farm. Another refugee from neighboring West Virginia, he'd once run bars and restaurants and farmed on the side, growing hundreds of acres of corn and soybeans and tending beef cattle. "I drank then," he'd said. "The whiskey kept me going."

Now Ernie spun me to face the window. He circled me, waving a black comb in his left hand, his razor glinting from the other; his thick black hair, oiled and combed back, bounced, and a lock fell across his forehead. "If you get Lost Valley," he said, "we'll be neighbors. I live just up the road in the white house."

I'd seen it: a sagging clapboard farmhouse, chipboard nailed over one window, its tin roof blistered with rust, its front porch jammed with old appliances, a car with a peeling blue vinyl roof parked in front; on the yard's edge a wooden barn had collapsed inward on itself.

"You'll farm the government more than anything," Ernie went on. "The paperwork alone will kill you. Now I've got all my land in the set-aside program. I grow weeds and briars, and they pay me ninety dollars an acre. That's more than I earned growing corn! But you got to tolerate their bullshit. Bureaucrats tell me when I can *mow*. I used to keep a billy club behind my bar for guys like that. I'd break their heads wide open."

"I hear you, Ernie," I said, scrambling for neutral ground. He made me bashful about my desire to farm. As his razor tugged at my hair, I felt myself clinging to my insider's knowledge, the monthly sermons in my new fantasy provider, an esoteric Mississippi magazine called the *Stockman Grassfarmer*, which railed against America's crazy system—low prices, narrow margins, government programs—that inevitably whipped conventional farmers.

Now Ernie was talking about his black Angus herd. One cold morning before his day job, as he lugged a hay bale through deep snow to a feed bunk in the corral, the greedy cows mobbed the forage and knocked him around like a pinball. "I got back to the barn, sat down, and cried," he said.

Eventually what he was paid for steers fell below his costs to raise them. "Damn Democrats. Jimmy Carter flooded the market with cheap foreign beef," Ernie said, jutting out his chin and jabbing his comb at the ceiling as he spun my chair to face the mirrors above the white Formica counter along the back wall. "Across the street, students were eating steaks from Argentina!" He swept his razor back toward the window. "I shot every one of my cows."

Jesus. I shook my head in sympathy. I imagined Ernie's dark eyes looking down a rifle barrel, an empty bottle at his feet. I knew the *Grassfarmer's* take on the arcane world of commodities and federal policy: bypass it by selling directly to consumers, or at least lower your costs through pasturing; even in winter, have the livestock harvest their own feed by stockpiling ungrazed grass.

Instead, in Ernie's experience, farming was a traumatic way to go slowly broke. His story sounded too much like my father's. I was eager to scoot off the green vinyl seat and flee.

"Well, Lost Valley will be a good investment if nothing else," Ernie was saying. "God isn't making any more land." He patted the sides of my head, his sad eyes meeting mine in the mirror. "Your money will be safe," he said.

I dropped my gaze, shamed by his subtext: *You're a cautious rich man, unlike me*. Then I felt a flash of irritation—as Ernie knew, I'd plunged into the local farming scene, if not fearlessly then at least with sincerity and goodwill. I'd toured farms with the county agent's "grazing council," a support group in recognition of the fact that Appalachia's slopes had eroded disastrously when plowed for corn yet were perfectly suited to raising livestock on grass. And not long after our arrival last summer, I'd begun milking cows at a small dairy farm a couple of times a week after work; lately I'd been learning about sheep by helping a local shepherd.

But as I scurried back to the manuscripts in my office, bespectacled and bookish in my blue Oxford buttondown, I felt very suburban. My father's words rang in my ears as I crossed the wintry campus: "The only way to make money in farming is if you inherit your land."

I'm scared, I realized. *Weird. No one is pushing me to farm*. No one, except myself. I found it unthinkable to change course, as if I'd be prosecuted for abandoning an old dream I'd revived. Exactly whom, I wondered, would I be letting down?

CHAPTER TWO

Molly is suspicious from a calfhood trauma: a guy tried to rope her.
—Farm Diary

THAT FIRST WINTER, FRIENDS BACK IN INDIANA WOULD WARN US when our former Hoosier hometown had been thrashed by wind and water and snow. "Watch out," they'd say, "there's a bad storm coming your way." On television the Weather Channel's radar confirmed this. A grainy mass of swirls was leaving southern Indiana and was bound for southern Ohio, coming right at us. But Appalachia's uplifted terrain pushed back against fronts driven by winds from the south and west, and urged tempests along an easier path. Just before hitting us, storms would turn north and punch Columbus. I pictured the heavy pressure ahead of a storm meeting the hills, slowing and filling space, climbing like water rising against a dam, and backing toward the approaching turbulence—the storm's own force an invisible barrier against itself.

The foothills were becalmed in summer. A maverick breeze might ruffle soybean fields in the bottoms, but hot gales from the plains that had scoured Indiana without resistance got confused when they met the cool, damp maze guarding that green kingdom. Gusts fractured into harmless puffs; in the valleys lay a stillness. Yet surprise abounded. The hills suddenly revealed secrets—or jealously concealed them in their folds; a casual visitor might never know that behind the dark ridge in front of him a valley stretched out in the sun.

In Appalachian Ohio, undulating hills jut three hundred feet above the flatter ground of the Allegheny Plateau, and form low ranges that curl protectively around valleys. For all the steepness of their ascent, the lush hilltops are comically rounded: mounds of emerald clay shaped by a laughing child. White mists hang above the wooded ridges after showers, and mist rises like plumes of steam off their wet green flanks. "Look," the people tell their children, "the groundhogs are makin' coffee."

Waves of settlers before us had flowed into this furrowed terrain, scouting territory like hens looking for safe nests. First came indentured servants fleeing Pennsylvania; then Scotch-Irish southerners, those too poor to afford land and too proud to work on plantations, made their way up the shaded spine of the Appalachians; and endlessly ever after came West Virginia's stream of refugees. They found niches. Hard against the gentle hills, they could see something coming long before it saw them.

Winter was descending. It was my last milking of the season, Friday, December 13, at Diana Carpenter's dairy and bed and breakfast, near Amesville on the county's northern edge. A beefy black-and-white Holstein cross named Molly, and a petite fawn-colored Jersey named Charlene were the first to be milked.

Back in August, Tom and I had arrived almost in time to see Molly give birth—and did watch her eat her afterbirth. Tom's eyes were huge, and I think mine were too, as we saw her chew and swallow a ropy mass of bluish, bloody placenta. Charlene also had a history with me: in September she'd tried to kick me in the head, her filthy hoof whizzing past my ear as I squatted beside her udder.

Now I was alone with this mismatched pair. Diana was next door in the tight room that received the milk; it was jammed with a stainless-steel tank, pipes, motors, and drain cocks, an array as confusing to me as the innards of a submarine. For milking cows, Diana had an old-fashioned flat parlor, meaning there wasn't a convenient pit for me to stand in while reaching up to attach the milking apparatus. I gave each cow a scoop of grain in troughs against one wall and slid bars behind their heads to keep them from backing away.

First you washed their teats—not "nipples," as Diana had corrected me tartly, amused. She'd warned me that getting kicked was a possibility. "You don't want to get your nose taken off," she'd said with a dark look. Keeping close to the cow was important: you're in a relationship with an animal that weighs around 1,000 pounds, and if it turns violent you'd be safer to be touching; that way, a cow's kick would push you away instead of striking you like a piston. I jammed my left hip against Charlene's right flank, bent from the waist, cleaned her udder with a soapy wet rag, and attached the claws: four stainless-steel suction tubes, one for each teat. I rose and sighed, always tense after Charlene's near miss. I surmised that her kick was meant to put me in my place, probably behind her and Diana.

After her kick, Charlene had glanced at me and tossed her head, turning back to the food I'd provided. The malicious gleam in her eyes had startled me—her fleeting expression appeared so self-satisfied, in its smarmy, gratuitous meanness so *human*—and I'd felt a flush of anger and wanted to kick her belly. Instead I yelled, "Bad cow!" And felt silly. Diana had shrugged, saying Charlene had tried to kick her son, and he *had* hit her, with a stick.

Tonight I removed the claws without incident, dipped Charlene's teats in iodine, and released her and then Molly. In the past four months I had learned to perform the entire milking and cleanup routine.

Having animals from which you harvested a regular product—like laying hens—appealed to me, partly because the animal got to live. And the demanding work promised cash flow, monthly checks for milk. There was an ongoing relationship and the comfort of daily routine. Dairying is all about routine, the cleansing purity of elemental work proceeding through rain and shine, through shifting human and animal moods and afflictions.

After I sent the two cows into an adjoining room for their post-milking hay, it surprised me to see Molly butt Charlene and chase her around the long wooden manger. When Diana returned from the milk house, I was grinning as I released another pair into the room.

"What?" she asked. She was a tiny woman of about fifty, with short brunette hair, lively brown eyes, and a flashing smile. Tonight she wore a red plaid flannel shirt, and her blue jeans were tucked into green rubber boots.

"Is *Molly* the boss of Charlene?" I asked.

"Yes. Molly's top cow. Charlene wants to get in first, with Molly, but she's unwilling to take the responsibility that goes with it. To be alone with Molly."

"Does Molly do that with the other cows?"

"No, just Charlene," she said, walking to the Dutch door that opened into the hay room. Through the door's open top half, Diana looked at the pair—one wide and fat-cushioned, one narrow and bone-sharpened. "But see how Molly blocks the manger?" Diana asked. "She doesn't want any other cow near her."

Soon after I started helping Diana, Kathy had surprised me one night by asking about buying her place. We'd heard that Diana had recently advertised the farm in bed-and-breakfast magazines. I loved her saltbox farmhouse with its simple foursquare layout of large rooms and gracious wood floors, though a two-story addition for boarders felt cramped and cheap. And it was far from town, way out in the boonies.

"Have you talked to her yet about it?" Kathy asked as we cleared the table before I left to milk.

"About what?"

"You know, her average occupancy, projected income. The length of the season."

I looked at her. She'd actually consider running a small Appalachian hotel even as she led a university college and mothered Claire and Tom. Yet I knew that Kathy's gift was seeing opportunity. Truth was, I assumed that Diana might sell her herd with the place, and the notion of becoming a dairyman in one stroke had thrilled and terrified me.

"Kathy, we can't," I said. "We'll fall apart."

"From what you've said, Diana seems to do just fine. We could consider it."

"She started the B&B first and grew into the dairy. She's juggling *a lot*."

I tried to picture us as professional hosts. We'd never even stayed at a bed and breakfast. Between us, only Kathy had acceptable cooking skills. Plus she was better with household chores. And better with people. I was in awe of her multitasking abilities, but I envisioned piles of dirty linens and endless strangers at our table as I obsessed over learning to run a dairy. I tried to imagine going through life chained to udders. The dark side of Kathy's competence was that she took on too much. She'd grown up as her father's main farm helper as he'd juggled three occupations; being stretched thin was normal for her.

Anyway, Diana soon sold her place to the daughter of a friend, as a hobby farm. And she told me she was taking her cows. She'd grown as a farmer and was ready to manage a bigger herd. For that, she needed more land. Her useable acreage wasn't much more than Lost Valley's, and a neighbor's house on the next hilltop loomed over Diana's valley, undermining her comfort.

In hill country, farmers get more property than they pay for—the slopes, if flattened, would bulge past the boundary fences—but the terrain also foreshortens distances: the ridge tops on either side of a valley move together, like cows crowding a trough. A neighbor standing innocently in his yard on the next hill appears to monitor your trek across your pasture. You want to get under the cover of trees on the field's edge, as jumpy as a rabbit.

There was a larger issue than whether Diana's particular house, land, and enterprise had been suitable: my growing ambivalence about cows. Dairying required a big investment, tied you down, and might get you kicked in the head. From boyhood, cows had always impressed me as beautiful—beef cattle, anyway, had seemed pleasingly sturdy, simple, sleek-coated creatures—but I'd never worked with them. Now I'd seen they were hard on fences, barns, land. Their manure was copious and hard to handle in barns and corrals, where the heavy mud-and-manure slurry could suck boots right off your feet. "A Holstein cow is the only animal that shits more than it eats," an ex-dairyman said when I'd griped about Diana's herd. Funny, but literally true, since a cow's manure is mostly water.

I'd become conscious as well of my aversion, which was mysteriously strong, to bare soil and erosion. Cows chopped unsightly trails with their big cloven hooves, and water sliced into the cow paths, turning them to ruts that ran with water that cut even deeper. On hills, as they spiraled around the slopes to ease their climb, cows gouged terraces as their hooves sent precious topsoil into the valleys.

Below Diana's house was a little pond, and one day her cows, trudging its borders when the soil was damp, had punched its gentle grassy banks into jagged stalagmites of greasy mud. I saw that the tensions between living in the country and farming lay right there, in that once-charming pond whose flat-topped dam, in a pinch, had made a handy lane to bring the herd in for milking.

With most of the milked herd lined in a tight row along the manger, I saw Charlene and Molly eating peacefully together. Svelte Charlene even pulled hay from burly Molly's mouth as Molly chewed hanks of the fodder.

"Look at that! Why is Molly letting her do that?" I asked.

"It's this frenzy they get into, afraid they're not going to have enough food. Then they see they do . . ."

As I finished milking, a cow named Alice, with bowel-control issues, shat in the parlor before I could get her out of the stanchion. This made me anxious—Diana had once hit Alice with the flat tines of a pitchfork for doing that, and Diana could be brisk with me. She just smiled, remarking that now Alice had *her* trained to open the stanchion faster.

With her cows all bred to calve at the same time, in harmony with spring grass, the entire herd's milk flow was tapering off. The cows were putting their energy into growing fetuses. In four days a trucker would collect Diana's milk for the last time until grazing resumed. She'd then hand-milk for several days and give the milk to Amish neighbors. Lactation would resume around March 15, when her cows would begin to calve. It seemed a long way off, the mud and wind of March, new life coming on in the cold rain.

We walked into the dark evening, mild for December and breezy. A square of yellow light from the milk house fell onto the gravel farmyard. We chatted about Diana's new farm in northern Ohio, about the new farm Kathy and I would own in spring.

"But it'll take a lot of work," I said. "Starting over here has been hard."

"Who knows," she said, "you might not last."

Her private thoughts had escaped in a flash from her keen face, still brown from summer, and then fled into the dusk. I felt like she'd slapped me; I was as surprised as when Charlene had lashed out at me with a dirty hoof.

"*Diana*," I said, drawing out her name, my voice as reproachful as I could make it. She averted her eyes, fell silent in the gloom.

But she was right. Maybe she sensed the angst beneath my cheery pose. Who doubted my future as a farmer more than I? Who was more uncertain about Appalachian Ohio? My latest complaint was that living here seemed to hold an increased mortality risk. From finding competent medical care to following the directions of distracted highway flagmen, you were on your own. I'd been unnerved one recent morning when driving to work across an icy section of the bypass around town: cars were sliding off the highway or rear-ending each other, but the radio was silent about the danger. Where was the warning I'd been trained by our station back in Indiana to expect? Why hadn't the police informed the radio stations? Why hadn't the radio staff bothered to call the highway patrol for road conditions?

Now, as I drove back to Athens over the twisty roads, I couldn't shake a callow but genuinely despairing feeling: *We shouldn't have to start over in such an ornery place.* A cow's hide is over half an inch thick; mine seemed considerably thinner.

And yet, although my course didn't seem simple, easy, practical—or even clear—I still felt a quiet faith in the strength of my dream. On a spiritual quest after graduating from the University of Florida, I'd read some of Søren Kierkegaard's writings and again comforted myself with one of his phrases, which had emboldened me to court Kathy at Ohio State: "Desire is a very sophisticated emotion."

Looking back, Claire's favorite aphorism, which she'd garbled from an old precept, also fit: "If at first you don't exceed . . ."

Claire and Tom accompanied me on Sunday afternoon to visit some of our chickens. Say what you will about this backwater, I thought as I steered my white Ford pickup on a southwesterly course out of town, you found a place to board excess fowl. I'd hauled a movable pen, along with about a dozen hens and our noisy rooster, to Willy Blosser's homestead. The rest of our flock clucked from makeshift quarters in an old plywood garden shed uphill from our house.

As we traveled along the aptly named Pleasant Hill Road, I admired rolling pastures and tidy homesteads. On scruffier Rainbow Lake Road, I pulled onto the gravel beside Willy's lodgings, a square structure of rough-sawn wooden clapboards painted electric blue. We got out and I reached into the truck's bed for a plastic ice cream bucket containing food scraps. Our chicken landlord's puppies toddled toward us from the darkness beneath his house, and Claire ran for them, her long brown hair shining in the weak December sun. Tom trotted behind.

Pulling on a wire loop, I dragged my pen to a clean patch of Willy's dormant grass. Back in Indiana, I'd built the "chicken tractor," covering its front with welded-wire and its back with roofing tin. I'd overbuilt it, using too many wooden braces, so to help move the six-foot by eight-foot coop across our lawn I had bolted lawnmower wheels to its base.

Claire and Tom were laughing at the puppies when Willy, slight and moon-faced, ambled outside, his German shepherd bitch slinking at his heels. I dropped our kitchen leftovers through the wire, and our buff-colored hens dove for the treats, overseen by their cackling red-hackled mate.

"Nothing's better for layers than table scraps," said Willy, grinning with approval. His dog, the puppies' mother judging from the swollen appearance of her teats, sniffed at the pen and avoided me. Willy stood with his left arm reaching across his body, holding his withered right arm against his side. He'd told me a motorcycle wreck had torn off the arm. Surgeons had reattached the limb, but it was without muscle, all sinew and bone, and appeared useless, a reddened stick with a hand. I was paying him $30 a month—an outrageously high sum I'd named—to keep our birds and to ward off the myriad predators that love to feast on chicken. I drove out every afternoon to feed and water them.

One day at the barbershop, Jim had recommended Willy when I'd asked about boarding chickens. "He's got empty pens from when he raised game-cocks," Jim said. "He used to take me to the chicken fights. We all got robbed one night at a fight. Three guys with bandannas tied over their faces held a gun on us and took our billfolds. I was so scared I thought I was going to shit out my own liver. The law started busting fights after that, and Willie quit."

Willy was, according to Jim, the one good egg from a nest of rural ruffians. "His brothers, Seth and Gordon, are real bad news," Jim said. "Thieves. Drug dealers. Before Willy got his place and moved out, they'd take advantage of his arm and beat him, take his money. He was always working at odd jobs, trying to get ahead, even though he's in pain all the time."

Claire swayed toward me with a large black puppy cradled in her arms. "I like this one," she said, her brown eyes pleading. "He's so *sweet*. I've named him Raffy. Can we keep him?" The pup gave me an irresistibly waggish glance and resumed gnawing on Claire's forearm. She'd inherited my love for animals, and though my own gift for naming them had faded in adulthood, Claire's was still flaring like poetry. Tom, his baby-fat cheeks flushed, his large blue eyes watchful, stood beside her, ready to go home but uncomplaining if a puppy were in the offing.

"Just three left," Willy said. "They're pretty much weaned."

"I'm sorry, honey," I said, "but we've got our hands full right now." The kids marched back to release Raffy to the dirt beneath Willy's house. Our middle-aged collie, Doty, wasn't very playful anymore, and I made a note to myself to get Claire a puppy as soon as I could.

Willy leaned on my rolling pen and inspected our "Easter Egg" fowl, fluffy chickens with feathered ruffs at their throats that gave them an owlish look. Their charm and their mysterious genetics—the unknown source of that blue-egg gene—had reignited in me a passion for chickens and for selective breeding that had lain dormant since boyhood, when hens and ducks had been the only farm species I could keep in our beach town.

Down the slope of Willy's brown winter-dormant yard was a graying chipboard shed flanked by small empty coops with rusty wire, infrastructure for breeding, raising, and separating pugnacious gamecocks. Suddenly our Ameracaunas looked as suburban and as foreign to this place as I felt.

"They're *cute*," Willy proclaimed, seeing nothing odd in a stranger's desire to board exotic chickens.

Driving out to tend our chickens every afternoon, and running up to Diana's a couple of times a week after dinner to milk her cows, sometimes I feared that I'd lost my sense of priorities. I pushed ahead, energized by the chance to make my dream a reality. But in our Bloomington years, I saw, I'd become oblivious to how myriad decisions and steady daily work and minor windfalls accrue and hold a body in place. And our move had bared a deeper conundrum, life's contingent nature: *this* depends on *that*. Just because we'd built a happy life once didn't mean we could again. I'd forgotten the effort it had taken to locate and create our showplace in Bloomington.

When Kathy was pregnant with Tom, we'd begun looking for a larger house in the country, yet our desire to keep our children in Bloomington's esteemed schools kept drawing us closer to town. After a year, when we looked at a featureless soybean field on the eastern outskirts we knew we'd found our plot. The square parcel on a narrow country lane was only a mile from the university and two miles from the downtown square. We'd have to build a house, but we might be able to afford to do that in a few years. Kathy convinced the owner to price the eighteen acres as two nine-acre rectangles so we could buy one. I keenly wanted it *all*, though Kathy didn't see why we needed that much land. Her logic prevailed: we got nine acres. Thanks to her, we'd always been frugal, banking every penny of my part-time teaching money and whatever else we could save from our jobs, and we'd be shrewd in this investment.

Still, even the land's owner was perplexed by our purchase when I told him we didn't plan to build right away. "Why do you want to mow weeds?" he asked me at the closing. "Because," I said, "it won't be available later."

Indeed, that next spring a developer began building Devonshire, an upscale "equestrian community," right across Russell Road, doubling the value of our $30,000 plot. We decided where our eventual house would go, and I laid out a graceful S-shaped gravel driveway right up to our imaginary garage door. I ordered pines for a windbreak to the west and north of our future home; taking turns carrying Claire in a backpack, we planted three hundred seedlings, lost on the edges of the open field. Every weekend that summer I placed four garbage cans in the bed of my little Mazda pickup, filled them with water from a garden hose, and drove down the bypass to give each tree a drink, sloshing all the way.

My father was selling his retirement business, a nursery, that year and gave me his diesel tractor, a miniature thirteen-horsepower Kubota. I towed it from Florida to Indiana on a U-Haul trailer and kept it in a concrete-block storage unit near our property. After the sun was up on weekend mornings, I'd drive from our house to the facility, climb aboard the little orange machine, accelerate it to top speed—eight miles an hour—and head down the blacktop, emergency lights blinking. Arriving, I felt intimidated by the size of our field. As foretold by its previous owner, the nine acres, freed from domination by herbicides and soybeans, had erupted in weeds: thistles, Johnson grass, ironweed, and giant ragweed reached eight feet tall, towering above me on the toylike tractor. From the shredded vegetation in a damp swale in the middle of our land rose the minty scent of pennyroyal—*pennyrile*, the Hoosiers said.

Sometimes our big meadow's wildness exhilarated me, but with my beach-town aesthetic, more often I felt awed by the power of plants, and a slave to chlorophyll. "How long does it take you to mow the place?" a friend asked, thinking in terms of hours or days. "Months," I said of my endless cycle. "All summer."

That fall, Kathy inherited $10,000 from her father's estate, and by the next summer, with Kathy expecting again, we decided to use her modest nest egg to turn our land's low spot into a pond—a centerpiece for our homestead, and that much less to mow. "Will it hold water?" I asked our excavator, Walt Taylor, who'd built a pond for one of Kathy's colleagues. Munching a Little Debbie snack cake, he regarded the declivity. "It will be a bad advertisement for me if it doesn't," he said.

A few broiling days later, an acre of raw dirt and a tall dam bisected our land. I raced around seeding the slopes and scattering straw to prevent runoff, until my back went into painful spasms. Turns out, I had inherited something too: my father's bad back. (My anxiety seemed mine alone.) Which was why I was lying on our couch, watching Claire play with our aging Labrador, Tess, when our land's closest neighbor—a scrawny, sharp-faced native called Termite—spied Kathy, seven months pregnant, carrying rocks off the pond's banks at my urgent orders. It was a scorching Saturday in July, 95 degrees and 100 percent humidity, and she was juggling hot stones the size of casserole dishes. Dressed in white running shorts and blue top, Kathy was filthy; sweat cut rivulets through the dust and chaff that coated her bulging belly. She looked nothing like what she was, a tenured professor and chairwoman of an academic department. Termite, wanting a closer look, drove over in his pickup, parked at the end of our driveway, and watched her stagger across the parched dirt. Finally he walked down. "Woman, what in God's name are you doing?" he asked.

She'd earned Termite's undying admiration—and so had I. "You married a horse," he told me later.

After we built and moved into our house, I began writing a gardening column, growing perennial flowers and vegetables, and fantasizing about small-scale farming. I was spurred to raise broiler chickens by the ideas of Virginia farmer Joel Salatin, a charismatic ecological-farming innovator. On his Polyface Farm, pastured meat chickens were the cornerstone enterprise. "Polyface is the prototype farm of the 21st century," an article in the *Stockman Grassfarmer* crowed. "This is how we should be farming and probably will be."

At Salatin's farm, I read, cattle grazed first, shortening the grass for laying hens, which ranged outward from a portable coop and ate fly larvae in the bovine manure; broilers followed this parade in floorless pens that were dragged across the pasture's tenderest regrowth. The meat chickens were ridiculously fast-growing—ready to butcher in six to eight weeks—so multiple batches could be raised throughout the summer to supply vital cash flow. And Salatin's holistic model was a humane alternative to filthy feedlots and poultry houses; although the birds' lives were brief, they were natural—they got to be chickens.

And I loved chickens. I was dangerously close to becoming a poultry nut who collected fancy fowl in bare dirt yards. Behind our new house, I had tried out several breeds, including the blue-eggers, and had bought Claire some colorful bantams. Salatin's use of an unromantic industrial strain, the husky, white-feathered Rock-Cornish hybrid, didn't fire my imagination.

Gradually, with the *Grassfarmer* stimulating memories of my father's pastoral farming and echoing Bromfield's vision of sustainable agrarian abundance, I saw the practical beauty of Salatin's choices, which culminated on the dinner plate with healthful meat. And I saw the sense in using a breed with wide consumer acceptance. Unlike hardier traditional breeds like Rhode Island reds, the absurdly meaty Rock-Cornish is double breasted. To modern consumers, *that's* a chicken, not some scrawny hatchet-breasted thing.

Our second summer in our new house, I dragged Kathy, four-year-old Claire, and two-year-old Tom to a Virginia motel with a cloudy swimming pool so that early the next morning we could join four hundred people for a day of touring Polyface and listening to Salatin. The farm in Virginia's bucolic big valley was beautiful, the activity there dizzying. At lunch we sat on hay bales and ate delicious Polyface chicken, basted in a neighbor's secret sauce.

"The market for broilers," Salatin said, "is insatiable."

I was hooked. I already knew how to raise poultry, and after Salatin's field day I went home and adopted his method, guided by his photocopied *Pastured Poultry Manual.* His instructions were always relentlessly practical—*this is the narrow gate you must pass through to be profit*able—but carried a mystical sense of being in tune with nature, hitching a free ride on all that solar power beating down. There was even romance in the flecks of dark green kelp meal that spiced his mash formula and made it smell like the sea.

I squatted in our new two-car garage with piles of wood and rolls of wire and built three rolling pens. Around our homestead I wore a blue nylon Polyface cap emblazoned with the "Farm of Many Faces" logo: an outline of a tree. Inside the tree were squiggles like leaves that formed the silhouette of a cow's head; inside the cow's profile was a chicken's head; at the center, a leaping fish represented aquaculture and the Salatin family's evangelical Christianity.

Soon I realized the unforgiving Old Testament nature of Salatin's commandments when I tried faithfully to follow them with my first batch of peeping yellow chicks. Cold, rainy weather set in that April just after I'd moved them from their basement brooder to our lawn. I ran an extension cord out to the pen for a warming infrared light under the roofed part, and tossed in straw so the bedraggled chicks could break contact with the wet ground. Icy rain dripped down the back of my neck as I endured a miserable full-immersion baptism as a farmer.

I kept the birds parked for three days until the storm passed. By the weekend, I saw bright red gobs of plasma in their droppings. I asked Salatin for advice in a letter—he didn't have e-mail yet—and by the time he responded ("You must keep them moving, no matter what") the chicks were shitting pure blood. With their diet of drug-free feed, the only protection the birds had from their intrinsic enemy, a tiny intestinal pest called coccidia, was to outrun it by being moved away from their own droppings, festering with the parasite's microscopic eggs. I treated the chickens with a powerful sulfa drug so their ragged intestines could heal. They survived, but this raised a paradox I couldn't resolve: the farmer's heroic struggle to nurture animals he'd soon kill.

Kathy was a good sport about helping me butcher batches of fifty broilers each summer. Friends loved the birds we gave them, and I sold some to a few grateful customers across the road in Devonshire. Meatier than store-bought birds (because I let them live a few weeks longer), their flesh was also firmer from their more active lives.

Claire and Tom at first were fascinated by the gore in their Dad's new hobby, but soon Claire disapproved. From the family-room window she scowled at my makeshift shade-tree abattoir off the back deck. Even to me, the death-dealing seemed endless.

Rather late I did the math: to earn Salatin's promised $25,000, a farmer netting four dollars a bird had to raise and sell 6,250 broilers. Which meant butchering more than 1,000 chickens each month in the growing season. And that meant spending your weekends cutting throats, shoving wads of scalding wet feathers into paper sacks, and tugging hot entrails from body cavities. "It's yucky," Salatin said. "It takes character."

It was too much for me.

I was slow to admit that, and felt relief when our move conveniently derailed my sideline. Raising broilers, aside from launching my journey as a farmer, made me realize how much I wanted a grass-eater to replace mowing. I had to mow constantly to keep the grass tender for those chickens;

after an hour I'd get bored, and even then, there was too much else to do. I began to see the beauty of grazing animals, which can harvest their own cheap, abundant feed.

But now, in Athens, my vision of ruminant-based pastoral farming sometimes felt like an obsession. And I had to admit to myself that it was keeping us unsettled. Our house was chosen as a short-term solution, and we'd never lived that way; it needed repairs and remodeling, but we were reluctant to put more money into it since we'd be leaving for a farm. Our Bloomington years seemed to be taking on a mythic quality for Claire and Tom, who spent hours watching videos we'd made of their cute infant selves and raucous birthday parties.

In Bloomington, Claire and Tom's bedrooms had overlooked the pond, an acre of shimmering blue water overhung by weeping willows and dappled with lily pads. Here, Claire's rusty casement window opened onto a narrow concrete back porch, and Tom's window faced a galvanized underground well. They had their own TV room outside their bedrooms in the walkout basement, but with the master bedroom on the main level, the house's layout was another foe to family togetherness. Claire seemed unusually short-tempered; Tom had gone almost silent. When we let the kids pick fresh wall colors for their new rooms, he asked for black; we compromised by buying him black carpeting.

On Monday morning I was writing a press release about one of our new books. About the time the kids started school the previous September, I'd started work. I went in early, at my desk in Scott Quadrangle before seven, sorting through manuscripts and deciding how to pitch them. The press hadn't had a publicist, just an overworked marketing manager, and I was eager to demonstrate my skills and their payoff. Having worked at Indiana University Press, much larger and with long-established marketing infrastructure, I was confident I could do the job in the reduced hours I'd negotiated. Off at three o'clock, I could meet the bus or pick up the kids; and I was off entirely every Friday, my dedicated farming day.

The phone rang—not someone asking for a review copy, but our realtor, Brian Winesap.

He was blunt: "That place will attract developers."

"What do you think it will go for?"

"I think it could go as high as $80,000."

I almost gasped. Five thousand an acre? Top farmland was selling for three thousand.

Kathy and I discussed the figure that night, after the kids were in bed. We'd made money on the sale of our Indiana home and could afford to bid that high, although from the looks of the cabin on the property, we'd probably have to build a house as well. Agriculture could never pay for such land, of course, let alone finance a new house.

"I could learn to raise livestock there," I allowed. "What about the price?"

"It's probably worth that. Think about the places we've been looking at."

"We could buy a lot more land with that much money."

"It's perfect for Claire and Tom," Kathy said. "And it's on a *lake*."

We decided to sleep on our decision. But Lost Valley's careless beauty had stirred us. Some combination of terrain, trees, and human use had united on the old hill farm. We knew that such places are rare, that when a plot is developed or divided, its natural contours and breaks are destroyed. In Indiana we'd transformed our rectangle by sowing grass, building the pond, and planting thousands of trees, shrubs, and perennial flowers. All that had made it ours. Yet even as I'd poured my heart into that homestead I'd wondered if it could ever become a *place*, at least in my lifetime.

"What do you think?" I asked Kathy in the morning as we dressed for work.

Sitting on the edge of the bed, she zipped up her winter boots and looked at me. "Let's use Brian's figure."

"Offer eighty thousand dollars?"

"Yes. Plus $101. My father lost out once by offering even money."

"I doubt anyone's going to go as high as $80,000," I said.

"Plus $101," Kathy insisted.

We delivered our bid to Ted Foote's secretary, in person, on our lunch hour. Hearing our voices, Foote appeared at the door of his office. Portly and sandy-haired, a country lawyer in his proverbial red-leather lair, he sized us up in sly glances from under his amber eyelashes.

I asked him about the cattle we'd seen. "They belong to a fellow who lives around the corner," Ted answered. "He took care of the place for Mabel, watched over it, kept up the fences. He knows he's got to remove them, unless the new owner lets him keep them there, of course."

Then he said the winning bidder wouldn't be announced until mid-February. Two months more of limbo.

I took comfort in reliving Mom's story about Dad's epic search for land after they'd sold their unprofitable ranch in the high desert of southern California. First they'd moved to Melbourne Beach, Florida, and Dad went to work for an old flying buddy, commuting up Highway A1A to Patrick Air Force Base. It was 1956. Meg was two years old, I was one, and Mom was pregnant again. Dad's job was helping oversee the maintenance and protection of air force and NASA facilities that stretched from Cape Canaveral through the Caribbean.

Mail arrived regularly from Doane Agricultural Service, which Dad had paid to tell him the best place in the United States to raise cattle. One evening he came home and told Mom he'd resigned, that he wanted another ranch. Her response was, "Why don't we go to Mexico?" She knew Mexico had affordable land, and they loved the culture. But Dad had learned that forage was the cheapest feed for cattle, and plants grew best where it rained. Doane had informed him that the humid Southeast was the optimum location to ranch in America.

Dad's first trip that spring was a bust: he bid on a nice place in northern Georgia, but the deal fell through when he couldn't get a clear title because the owners were divorcing. Dad hit the road again two days after my brother David was born, in June 1957, and stopped for lunch at a diner in the southwest Georgia town of Albany. He'd been looking at the red clay and thinking it would hold water. He walked into a realty office across the street, and a man there told him that just to the north, outside the town of Leesburg, was a failed turkey farm for sale. Dad liked it, 850 level acres along the Muckalee Creek; it was called Stage Road Ranch, because an old coach road formed one border. He bought it—the little white house, red barn, neglected pastures, and mature pecan trees—for $69,000, a pittance compared with what he'd sunk into the California place.

Dad went to work improving Stage Road Ranch: he sowed orchardgrass in the shady pecan grove, where the northern species thrived in the cooler microclimate; he dynamited stumps in the pastures, bought earth-moving implements for his tractor, and dug a pond. Instead of buying registered cattle, as in California, he ran a herd of cheaper and hardier crossbred mother cows. He'd realized that he was paid for putting weight on cattle, not for their pedigrees, and he even bought dairy steers to grow out. This was unusual at the time; rangy Holsteins didn't bring a premium price at market, but they cost little to buy and made money if fed cheaply. He cut and hauled tons of grass, packed it into long trenches in the red earth, and made silage—fermented forage—for nutritious feed.

As always, Dad bought expensive new machinery. He had little aptitude for working on equipment, so buying used wasn't a consideration. Having also bought his land and livestock, he had to roll the dice on a big scale for cash flow; he needed big, steady sales. After three years, when he'd made enough silage to feed one million dollars' worth of cattle, Dad walked into a bank in Albany, confident he could get a loan for that amount.

Of course the banker said no to the crazy Yankee.

Dad's ranching career was over.

We left in January 1960. I was almost six, old enough to climb out of our vehicle and open gates for Dad. I understood the meaning of our last drive around the farm among the cattle in the pastures. I don't remember what was said to me, yet I knew we were losing our beloved home place. As I looked at the cattle and we said goodbye, I was overwhelmed by sadness.

I retain vivid images of Georgia, but have only jagged fragments of memories for several years after we moved into our concrete-block Florida house. Dad went to work again for his friend at Pan American, back in the workaday world, paying his dues. Our new hometown, the pastel-colored subdivision of Satellite Beach, had popped up like a psychedelic toadstool in the palmetto thickets south of Cocoa Beach.

Meg and I and two younger brothers, toddler David and newborn Pete, were now growing up in an earthly paradise. Satellite Beach was situated on a long narrow strip of sand, a barrier island between the Atlantic Ocean and the Indian River, a broad estuary. The breeze off the ocean a block east of our house cooled the humid days. When Mom drove us across the river on a clackety wooden bridge to buy luscious mango fruit, the river flashed and dimpled in the sunlight. We fished, swam, sailed, and water-skied. Clouds piled up like white mountains every afternoon and it rained. Pelicans glided over the beach and dived for fish in the surf. In winter, endless rafts of migrating wild ducks darkened the river's face.

But in family life, there are *before* and *after* divisions. California and Georgia were before. My parents settled into a typical middle-class, middle-aged life. The *Progressive Farmer* still arrived in the mailbox, but I don't recall Dad looking at it.

Mom told us vivid stories of another lost world, that of her small-town Oklahoma girlhood. She grew up in a large extended family that hunted and fished together, endured the Depression, were poor but didn't know it, and took in orphans and children from broken families. She recalled ranching in California: buying a prized Hereford bull, herding cattle from horseback, camping in the mountains, and fishing in Mexico. To soften the image of our silent, driven father, she told humorous stories about his learning to rope by using an irate billy goat, of overpaying his helpers in Georgia, and of locals' shock everywhere at encountering such an alien creature.

I helped Mom garden. Our efforts to raise vegetables were doomed— the soil lacked organic matter, and the humid air was full of salt spray and insects—though it seemed we tried every year. When Mom took us to visit her parents in Oklahoma, the chickens and ducks in their yard enthralled me. A black-and-white-barred Plymouth Rock hen drinking from a battered tin pan stopped me in wonder. Mom's father, Delbert Rounsaville, took me to his coops and pointed out the flashy red roosters, telling me their names—"Geronimo" and "Red Eagle"—which he probably made up on the spot for me. He wore string ties and smelled pleasantly of pipe smoke. On one leg he wore a tan elastic Ace bandage and once unwrapped it to show me a patch of black flesh where a water moccasin had bitten him as he waded a creek. Noting my careful movements around the fowl and how I picked up feathers, he saw a kindred soul.

Mom let me raise laying hens behind our house. We couldn't have a rooster because of the crowing; although I tried twice, a neighbor always complained. One year Dad bought me a trio of ducks, surprising me with his interest and enthusiasm. I hatched their eggs in a primitive incubator, and Mom crumbled the yolks of hard-boiled eggs for the ducklings' first food.

Someday, I vowed, I'd have lots of chickens—even crowing roosters—on my own farm.

When I was a teenager and Dad talked of buying a retirement place, of returning to the land one last time, I said, "I'll get my graph paper. Let's plan it!" He'd have another chance at farming, and would be talkative and have time for me, his ardent helper in achieving his lifelong desire. He only glanced at me and said bluntly, "You can't draw the layout before you see what's there." I'd imagined my own little Eden countless times, though, and I could see it.

"You won," Ted Foote said.

"Pardon me?" I leaned over the phone on the corner of my desk that Wednesday afternoon. Manuscripts were piled on either side of my computer. A raw February wind clawed at the double-hung window at my back in the brick dormitory where the press was housed.

"You're high bidders. Congratulations."

"Wow. That's great," I said. But I felt unsure—of course we'd won, bidding so much!

"Now, we can't close until April," Ted said. "We've got to make sure the title is clear, run notices to forestall protests from any unforeseen heirs. I don't expect problems, but we've got to jump through a few hoops."

"Okay," I said. "Can I go ahead and ask the caretaker to remove his cattle?"

"Sure. That's Massey Taylor. He's in the book. But speak loudly. He's hard of hearing. Deaf as a post, actually."

So, fifteen years after Kathy and I had met, just up the road at Ohio State, we'd acquired a farm in Ohio. With a trembling finger I punched in Kathy's office number to tell her the news.

CHAPTER THREE

I helped de-worm and treat for foot rot 200 sheep. You'd be amazed
how long three minutes takes when you are holding a struggling
ewe in a trough of caustic liquid.

—Letter to a friend

I ARRIVED ONE MILD AFTERNOON TO TEND OUR CHICKENS AT WILLY'S
and found my rolling hen house empty. Buff feathers littered the grass.

Willy came out of his blue house. "A dog got in," he said. "It went in
over the top and busted through the wire." He put his good hand on the
wire portion of the roof, and I saw where he'd stapled the wire carefully back
to a brace. Willy's German shepherd slunk around the corner of his house,
glanced at me, and disappeared beneath. Willy said, "I'm sure it was her.
She's a sneaky damn thing."

I stood staring at my silent pen. How could he have let this happen?
He knew that everything in the world liked to eat chicken. And that dogs
killed for sport. But he didn't raise chickens anymore. Even as I felt my
anger rise, I knew it was an owner's endless duty to protect his own poultry. Wild predators—raccoons, opossums, foxes, bobcats, coyotes, snakes,
minks, skunks, owls—owned the night. And in broad daylight, frenzied
dogs might massacre your flock. I'd learned this in Indiana when we started

47

losing chickens. One morning I saw a dog at the henhouse, grabbed my shotgun, and slipped down there. The dog and I met in surprise at the corner of the coop. The stocky brown Chesapeake, a gorgeous animal, moved aggressively toward me. I fired into the ground at his feet and he fled. I'd just buried Tess and was grieving. Although I couldn't bring myself to shoot the dog, I thought my shotgun blast had scared him off forever.

But a dog that's preyed on chickens, I'd learn, enjoys the experience too much to quit for long. One evening that summer I came home to find dead hens strewn across the lawn, and more mauled carcasses in weedy field borders and lodged in brush.

Willy and I stood staring into the empty pen. I guessed he'd removed the carnage to spare my feelings. I gripped its wire top, feeling ill, now angry with myself. Who ever heard of moving chickens? My novel solution of boarding them had had tragedy written on it from the start. I was a guy who'd showed up at Willy's with an odd contraption and a fistful of money. He'd been agreeable, but hadn't changed his life for me. He was a poor, disabled man with a few unfenced acres. Willy's mild, helpless attitude expressed his view: *Stuff happens. You can't fret about it, escape it, or cry when it does.*

But I wasn't Willy.

"If a lamb dies, just drag it out of the barn and leave it in the farmyard for me," Mike Guthrie had said. Thus I was apprehensive, driving to his farm that rainy Saturday morning. Since talking to him at a pasture walk, I'd apprenticed myself to him to learn how to raise sheep. Or anyway, like Diana, he'd agreed to let me help him.

Towering over me at my father's height, six-foot-two, Mike wore his long reddish hair pulled back in a ponytail. His pale blue eyes peered at the world through the owlish lenses of gold-framed granny glasses.

Mike despised mainstream agribusiness and had provided a welcome corrective as well to the *Grassfarmer's* ambitions for me. About that periodical, my bible: "It has an agenda, selling its philosophy and its experts, pushing that 'make a doctor's income from grass dairying' line. The editor makes easy money by selling books, workshops, and cassette tapes."

About full-time farming: "You can't expect to make a living from farming. Every 'full-time' farmer I know has outside income or a hidden trust fund." About my latest farming hero, Joel Salatin: "He's a freak. The exception that proves the rule."

Farm-sitting while Mike and his wife took a quick trip to Delaware to see his mother at New Year's, I felt unprepared. We'd done so little for his flock that I didn't have the confidence I'd gained under Diana's somewhat more gentle mentorship. I drove southeast away from Athens, passing the turnoff for Willy Blosser's, traveling down an old state highway that curved around steep hills and rock outcrops. Unhindered by traffic lights, I passed through the crossroads of Shade, turned right onto a narrow blacktop lane, and followed it to a potholed gravel road. In a mile I saw Mike's farmstead: a red shop and weathered farmhouse on a stony rise, the old barn below. I parked in front of his shop and got out, listening to the cry of a crow in the silence.

I walked downhill to the tin-roofed barn. It had been built a hundred or more years before at the bottom of the wet slope to take advantage of a natural spring for watering livestock. Mentally I moved the barn from its dark hollow to the top of the next hill, where the winter sun could at least reach it. When I'd made the mistake of telling Mike this idea—feasible in the age of piped water—he'd exploded: "Build a new barn! You can't make money in farming by *spending* money! That old barn is *free*."

As I entered the barn's dim interior, my nostrils tingled with the sharp ammonia smell of lambs' urine. Mike had brought in his lambs from pasture to eat hay and a little corn; he was holding them, hoping for a winter spike in prices. They scrambled away from me in silent flight, and I parted the flock neatly in two. At my feet in front of a hay manger, a white lamb the size of a large dog was stretched out dead on the strawy floor.

Mike had explained that lambs gorging on feed can get what shepherds call "overeating disease," an explosion of bacteria in the gut; it kills them quickly. There's a vaccine for the malady, but sometimes even inoculated lambs die, often the fastest-growing ones. This lamb looked like one of the biggest in the pen. A ninety-pounder, I thought as I grabbed one of its rear legs at the hock and leaned forward to drag it into the muddy farmyard. Another crow flew past, cawing.

The next day, I found another lamb dead; this one had jammed itself between a wall and an open gate. When Mike called to check in, I told him of the losses, and he reacted with grim satisfaction. "Haven't you heard?" he said. "A sheep is just an animal looking for a place to die."

No wonder some in the county agent's grazing group called Mike Guthrie "Mike Gruff" behind his back. Nobody could disparage his chosen species faster than he, and he loved to needle dairymen about cows. When I'd bragged about my milking: "Dairy isn't sustainable here, and not in most places without government support. Besides, dairymen have to put up with dumb, destructive *cows*."

For all of his provocation, we hit it off because we were alike in at least one respect. "For me," he'd said, "farming is romance, not business." Developing a farm that was sustainable in hill country was more important to him than trying to maximize profit. He sought a pastoral enterprise that would permit an Appalachian family to net $10,000 a year as one source of income.

"I believe in free sunshine that grows grass," he'd said one day in late summer as we were deworming his ewes. This involved jamming the nozzle of a chromed drench gun in their mouths and shooting down their throats nematode-killing medicine that looked like yellow Kool-Aid. Which they hated, so they thrashed and reared in Mike's homemade chute of poplar slats.

Leaning over a ewe and getting her in a headlock, Mike said, "Lots of farmers have gone broke producing bumper crops, doing everything right, following the latest advice." He straightened up and pierced me with his pale eyes. "Forget magic bullets. I believe in the wisdom of the farmer. But you've got to be patient. And you've got to think for yourself, Richard. I know what *I* want. And I don't want to work my ass off getting it."

Even with lambs in the barn, Mike's chores were simple compared with milking cows. Every day that week, I'd climb into his barn's dim hay mow and throw down a few bales to refill mangers; with an old coffee can, I scooped kernels of whole corn from a rusty fifty-gallon drum and trailed the slick grain into Mike's feeders, old boards knocked together into simple V-shaped troughs.

His two flocks of ewes, each escorted by preoccupied rams intent on the business of making April lambs, required no care other than breaking the ice a few times in their water troughs, old porcelain bathtubs. But they didn't seem to be drinking, and regarded me suspiciously from the yellowed grass that carpeted the sodden hills.

I drove everyone to Lost Valley one Saturday in February—even Doty, wagging her tail in back with the kids. Kathy had borrowed the cabin's key from Ted Foote. The weather had turned freakishly warm—a high of seventy-five forecast—and we felt energized. From a kid tape in the van's cassette deck, some man with a kindly voice warbled about moonbeams.

"Can we get a horse now?" Claire asked.

"Soon," I said. "We should probably be living out here."

"First," Kathy said, "we need to look around in the cabin and see how we can fix it up."

"That's Massey Taylor's place," I said as we passed a low-slung white ranch house on Ridge Road. I'd stopped by the previous week to ask him to remove his cattle and had found him sitting in the open door of his garage in a lawn chair, looking out. I'd had to yell until my jaw ached to make him hear me; Massey had kept mumbling about his hay baler that he stored at the farm. "It was strange that he could barely understand me," I said, "because he was listening to bluegrass music. The radio was on, anyway."

"He's a sponsor of that show," said Kathy, to whom the station nominally reported.

"Massey?"

"Yes. He's an expert on bluegrass. I've heard him call in and correct the announcer about when or where a song was recorded. He knows the names of all the musicians too."

I turned onto Snowden Road and in a moment parked in front of the cabin. The kids ran toward it as Doty sniffed fixedly at a spot at the edge of the farmyard. Kathy unlocked the cabin's back door, and we crowded into a small kitchen that opened into a sitting room, getting our first good look. The cabin was a dank time capsule. Red-checkered curtains framed the brown wooden windows; horses of black wrought iron pranced above the stone fireplace. In the other corner was a cramped bathroom. Upstairs, in the roof, were two tight bunkrooms and a window in the west gable overlooking the woods at the entrance to the farm.

"It's small, but maybe we can add on," Kathy said. She started to sit on the nubby burnt-orange couch in the living room, but then thought better of it. Claire and Tom left to explore the farmyard pond.

"Did you notice the bottom logs are rotting?" I said. "Someone built this without a foundation—just laid down logs on a layer of stones in the dirt."

When Kathy went to check on the kids, I hiked into the south pasture behind the cabin. After steeply rising, the land leveled into a grassy ridge, strikingly flat-topped; descending the other side, I angled toward the lake in the direction of birdsong, my pace quickening from the force I sensed: spring already, sopping up returning light and warmth, gathering its power. Trees had grown up along either side of the boundary fence, so I could glimpse only patches of water. In a clearing at the property's woodsy far corner I discovered a small pond. Or rather, the remains of one: the dam had been breached, a section in the center missing; a brook trickled through the opening toward the lake. I recalled how Massey's cattle had blurred the contours of another small pond in the north pasture, below the loafing shed, by wading and drinking. But this pond, from the looks of the sharp cuts in its dam, was destroyed intentionally.

I returned to close the cabin and decided to go inside. In the sitting room, I saw something shiny lying on the mantle. I picked it up, a Polaroid snapshot of a shaggy brown and white pony behind the stable on the cabin's east side. On the back of the photo, someone had scrawled, "Died in 1975." The animal must have been Kenneth and Mabel Vaught's. In the watery photograph, the corral leaned downhill and the pony sagged with age. The stable, the most ramshackle structure still standing on the farmstead, appeared past its prime even then, wearied by the drainage coming relentlessly at it from the hills above. I studied the image from twenty years before; it looked like the end of something.

I heard Kathy shouting in the farmyard and stepped outside. "Claire, take off those shoes before you get in the van!"

"Look at your shoes!" Tom said, following his big sister across the farmyard. His voice went shrill in joyful amazement and mock frustration. "*Oh,*

Claire." He laughed helplessly, his cheeks red, thrilled to his core by her latest escapade. Claire clomped toward the van, her white running shoes encased in greasy taupe mud so thick it appeared that she was wearing overshoes. She crouched and swung her elbows, scrunched her face at Tom as if she were a crone whose shoes weighed a thousand pounds. Tom held his stomach in pain now, and flipped forward at the waist, a boneless puppet, his hair fanning above the gravel. Doty, her white belly matted with dripping inky muck, flinched at the kids' hilarity. The ginger-colored dog ran to me and pressed her wet nose to my fingers for reassurance; a fusty benthic odor rose from her.

"Mom, look at Claire's *shoes*," Tom said, pointing. He was our neat child, the delighted and appalled witness to his big sister's regular plunges into anarchy.

Kathy, carrying his sweater and Claire's jacket, shook her head in brisk confirmation and looked at me. "Claire and Doty walked around the pond below the shed. Massey's cattle have really messed up its banks."

"Are his cattle still—"

"Yes," Kathy said. "They are."

I accelerated my research into pastoral agriculture, studying the *Grassfarmer* with renewed interest, and visited Internet sites devoted to farming. Every night before bed, I read shepherds' e-mails on Sheep-L, and pondered talk about grazing issues and forage species on Graze-L.

Mike Guthrie was too busy in his day job as a carpenter, and too far past his passionate early years of farming, to have much patience for my questions. But one late winter day I landed his story, and I learned that grass farming had been a redemptive climax in his farming career. We talked in his cabinetry shop beside his house as rain pinged on the tin roof. Growing up poor in Delaware, Mike had dreamed of a life on the land. He'd headed to southeastern Ohio after college, in June 1978, eighteen years ago. The region was growing in reputation as a countercultural mecca and had been endorsed by the *Mother Earth News* (which I also was reading at the time, down in Florida, Dad having learned about it in the *Wall Street Journal*).

Mike may have arrived in Appalachian Ohio looking like another hippie homesteader from the East, but I'd learned that he was driven by the most timeless and traditional of values—rural values. He honored "neighboring," passing time with those who lived around him, no matter how politically and socially different they were. And his flashing anger at human hypocrisy—at the *Grassfarmer*'s emphasis on income, at the way the area's few full-time farmers lorded it over everyone else at grazing group meetings, at America's indebted consumer society—hinted at something I took for a spiritual approach to life, though he didn't seem at all churchy.

At first he'd framed houses for area contractors, he told me. He and his wife found their hill farm—then overgrown, the house shabby—and got a deal. Over the years they'd sold beef, honey, eggs, vegetables, and maple syrup. Mike had even put in his own stint raising broilers on pasture. I got the sense that he and his wife, Carol, had worked themselves half to death with the *Mother Earth News*'s do-it-yourself diversity ideal. But in the past two decades, while producing much of their own food and earning a little money, they'd become farmers. (Carol was usually away when I visited their farm, working as a physical therapist in Athens; their daughter, Liza, was a high school senior whose dream, far from farming, was to become an interior designer.)

Mike said his breakthrough in redeeming his farm with grass came when he happened to hear University of Vermont agriculture professor Bill Murphy talking on an Athens radio show. This was in the mid-1980s, just before publication of Murphy's influential book about New Zealand's pasture-based farms, *Greener Pastures on Your Side of the Fence*. Murphy had learned that the Kiwis had adapted the ideas of a French biologist and chemist, André Voisin. Also a farmer, Voisin had observed his cows' behavior and studied how their grazing affected plants. I'd already read Voisin's 1959 opus, *Grass Productivity*, amused at the esoterica—Voisin tallied a foraging cow's chews per minute—and by his news that a cow will graze for only eight hours a day and then stop, whether she's gotten enough to eat or not.

Murphy took Voisin's forbidding thesis and put it into ordinary language for American farmers, bringing news of the latest agricultural revolution. Mike had listened intently as Murphy said that by rotating animals through small paddocks, instead of letting them roam large areas for long periods, they would harvest their feed with less being trampled and wasted. And plants, when given time to rest, grow nearly ten times faster than if they're bitten off each time they attempt to regrow leaves. Heavy grazing followed by rest mimics what occurs in nature when a mob of animals—bison, say—trashes a grassland but then moves on. And it results in diverse pastures filled with white clover, which concentrates protein and minerals.

Mike began buying movable electric fencing supplies and trying the Voisin-Murphy ideas, thus becoming one of the first farmers in America to attempt what has become known as management-intensive grazing. The hyphen in that term puts the emphasis on management because the farmer is in daily control, deciding where and for how long animals graze. The labor certainly was intensive for Mike. Lacking permanent fencing, he relied on flimsy but electrified wires, which he unrolled from plastic spools. At least once a day he moved his fence lines—fiberglass rods and three strands of wire—advancing his flock.

"It seemed like all I did all day was tear down and set up fence," he told me. "But it worked."

His battered farm began to grow lush forages, which flourished through the hot summer and into the dry fall because of healthier, deeper roots. Soon he could run more sheep and earn more money. He built permanent border fences and used his moveable electric lines for internal divisions until he could replace them.

"Make time your friend," he said, giving me a knowing look through the dusty lenses of his granny glasses. For his part, Mike said he was netting $80 a year from each of his ewes, despite the fact that his lambs, late born and pasture raised, never topped the market. Given the size of his flock, that meant he netted $12,000 to $15,000 a year. Not enough to live on by itself, though still impressive small-farm income in America.

The endless subtleties of grass farming that had galvanized Mike now fascinated me. In theory and, from what Mike said, sometimes in practice, pastoral farming forms an elegant whole. The animals, their reproductive cycles, the pastures, and the farmer's efforts move in tune with the seasons, with the entire tilting, spinning planet.

As a boy, reading Dad's old farm books, I couldn't tell sheep breeds apart; in photographs they'd all looked the same: white and wooly, a poor fantasy livestock. The real ones lived far from our Space Coast boomtown.

Studying the species in Athens, I learned that sheep and human fates have been entwined since prehistoric times. Humans made their significant transition from hunting to agriculture about ten thousand years ago with the taming of three ruminant species: sheep, goats, and cattle. These animals and their wild kin, not scythes or mowers, created the pastoral landscapes that humans seem instinctively to crave. Ruminants can convert cellulose—the cell walls of plants that make up much of the organic carbon on Earth—into milk, meat, cheese, leather, and wool. Humans can't digest much fiber, and neither can most animals. The ruminant is nature's original and unsurpassed creation for thriving upon this abundant resource. Flocks and herds, subsisting on nothing but plants, a little water, and a pinch of salt, built fortunes and empires over the epochs as they drifted across plains. Chickens and hogs, which need grain, a concentrated energy source, were kept in small barnyard populations.

The ruminant employs a big, complex digestive apparatus to make a living from cellulose. The sheep's small intestine alone can be ninety feet long. The rumen, or paunch, where forage is first deposited, is what makes all the difference. This stomach is a fermentation vat, which ranges in size from five to ten gallons in sheep. Along with the three other stomach compartments, the paunch fills a ruminant's left side and extends into the right. These are prey animals—heads down, grazing, they're vulnerable to attack—and so they swallow forage quickly and later burp it up from the rumen and chew it more thoroughly. While ruminating, they like to relax on a rise where they can see danger coming.

"There is an obvious sensual satisfaction to the sheep in this laborious business of cud chewing," observes Allan Fraser in *Sheep Husbandry and Diseases*, a 1957 Scottish treatise I mail-ordered from a dealer as I built my sheep library. "A flock ruminating and at rest on rich pasture is one of the most placid and contented spectacles a disturbed world presents."

Appalachia's foothills, now reforested, had once been grazed to their peaks by sheep. Ohio, and its southern region in particular, had been a major wool and lamb producer. As many sheep grazed Ohio during the Civil War as were now spread across the entire United States. Much of this was due to the demand for wool for clothing and military uniforms. Now, despite its huge cities and large population, despite the rise of synthetic fibers and the declining numbers of 4-H kids who exhibited animals, Ohio remained the state with the largest sheep flock east of the Mississippi.

As I began to consider raising sheep, I realized I'd be shifting from one despised species to another: sheep seemed to rank below broiler chickens (and even swine) in their public image. But I'd pretty much ruled out dairy cattle, and beef cattle weren't as profitable as sheep, from what I'd been hearing. Lamb, a delicacy, escaped the wild swings in the commodity meat market. Shepherds had only to master the two great challenges to the species: canine predators and internal parasites.

And sheep seemed a sensible size for a small farm and its humans. Even their manure was reasonable: small, dry pellets, like the droppings of big rabbits. I saw, too, that sheep were right for the hills. About the weight of deer, they didn't grind away at the slopes or create quagmires around watering holes. Their size made them the right scale as well for a middle-aged man with back problems that, if my father's history was any guide, would only worsen.

Rams could be dangerous, Mike had warned me before I farm-sat for him, some breeds more than others. But ewes were harmless creatures. They didn't seem to have human personality faults, like cows, but were more animal-like—comfortably alien beings. Their odd eyes—the color of mustard marbled with caramel, the pupils a black slash that rounded in dim light—regarded me without malice or calculation. Mike's tamer sheep showed mild interest: *Got anything for me, boss?* Those with wilder dispositions stared back in wide-eyed horror, heads lifted. Sheep seemed to have an emotional life; at least they experienced joy, running and leaping when happy. Sheep *didn't* all look alike, I was learning, and the temperaments of the different breeds varied within the admittedly narrow parameters of sheep personality.

In other words, and for many reasons, I'd discovered the hidden beauty of sheep. Physically humble creatures, they possessed a functional beauty, a rightness of being fitted to the land and its labor. Sheep weren't showy animals, but I saw they were suitable.

I'd found my species. A shepherd I would become.

On a cold Saturday in March, I drove out to Lost Valley at daybreak for some guerrilla frost-seeding of clover onto the farm's muddy cow paths and Massey Taylor's tractor ruts. With a red plastic hopper strapped to my chest, I turned a crank to send the round seeds bouncing across the ground and into the honeycomb of tiny holes heaved by frost. Then I stomped straw into the jagged wheel ruts leading uphill into the north pasture, and watched with consternation as Massey's cows appeared and began to eat the shiny golden roughage for breakfast. I tossed sticks and rocks at them and they slowly left.

When I walked toward the farmyard, eager to get home for my own break-fast, I was startled to see Massey's truck parked there. I hoped he hadn't seen me assaulting his cattle, and was relieved to spot him standing in the lane past the barn, studying his hay baler. He gazed at it as I walked up to him to assert my ownership—still not official, but surely he didn't know that.

"This is my baler," he said. "I've made a lot of hay here. Never seen the grass like this place grows."

"There's no hurry in getting it out of here," I said, "but—"

"What?"

"YOU CAN KEEP IT HERE," I yelled. "BUT REMOVE YOUR CATTLE." It was exhausting to shout, plus it felt hostile.

He sat down heavily on the baler's tongue, which was propped on a con-crete block, and rested thick hands on his knees. I noticed he was missing the ring finger on his right hand. Massey had milky blue eyes and crewcut gray hair. His shapeless build, square red face, and stiff denim coat brought to mind a Russian peasant; or maybe that sense arose from the stubbornly elliptical way he conversed.

"This place grows ironweed eight feet tall," he said. "Cows won't eat it. Or horses. Jams the baler, those stems."

"I guess you've been farming here a long time. What—"

"What's that?"

"WHAT HAPPENED TO THE POND IN THE SOUTH PASTURE?"

"Oh, had a cow get stuck, going for water. She died before I got there. I took my backhoe and just opened up that dam."

"Massey, I need you to remove your cattle."

"What?"

"REMOVE YOUR CATTLE!"

He continued to look at me without comprehension, or at least without expression. "What are you going to raise?" he asked.

"Well, for starters, sheep—"

"What?"

"SHEEP!"

He blinked and an impersonal sorrow suffused his features. "Folks have tried that," he said. "Dogs and coyotes kill them. Worms get them real sick too."

"There's new ways—"

"What?"

"I'M GOING TO TRY!"

"Sheep," he said. He looked at me sadly. "They'll eat the ironweed anyhow."

I wished I could tell Dad, gone ten years, about all I was learning, about the advances in grass farming—new theories, forage species, tools—since he'd been a pioneer in pastoral agriculture in California and Georgia. He'd used weak, clunky electric fences that constantly shorted out. Hence his desire to feed his cows efficiently on silage he made. His vision of paradise became vast hay and forage fields that surrounded a feedlot where his cattle awaited his deliveries.

I tried to imagine our having a talk about farming that went beyond a few sentences. He'd been conversant in more realms than I could imagine, but Dad was introverted and modest. He rarely talked about himself or mentioned his accomplishments.

His first passion, for aviation, was ignited at age seven when he listened to a radio broadcast of Charles Lindbergh's solo flight across the Atlantic. Dad took flying lessons as a teenager and, according to his pilot's log, made his first solo flight at the Detroit airport when he was barely eighteen. When he enrolled at Cornell University it was to study agriculture, but he still flew airplanes. When he landed one on a campus lawn, Cornell officials asked him to leave. It was 1938.

He headed for San Diego, for advanced pilot training at the new Ryan School of Aeronautics. The school was founded by T. Claude Ryan, who'd built the *Spirit of St. Louis*, the airplane that Lindbergh had flown in his 1927 transatlantic crossing. The Ryan school soon hired Dad as an instructor, and he flew his own airplane in a cross-country race. Later, when war came to Europe and Asia, he made regular flights across the Pacific, delivering B-24 bombers and Catalina seaplanes to Australia; for a time, he secretly trained British aviators at a school in North Carolina.

After the United States entered World War II, the Army Air Corps had him test prototype aircraft at Dayton, Ohio; by late in the war he was leading bombing raids into Japan. When Mom asked him why he never talked about the war, he said no one who had cleaned the bodies of a flight crew out of an airplane discussed it. Once, watching a war movie on TV with me, he mentioned that when someone got killed, or when an airplane simply didn't return from a mission, the aviators said, "He bought the farm."

Such was the promise of agriculture in the mid-1940s. Untold thousands of servicemen dreamed of starting a "chicken ranch," their fantasies fueled by the trauma of war and by books that portrayed the peace, security, and satisfaction of work on the land. Dad, dangerously thin during the war, a heavy smoker and drinker, was one of those dreamers.

Family lore has it that he was the first American to land an airplane in Tokyo after the war ended. He was twenty-six on August 28, 1945, when he flew in the general who'd directed the heavy bombing campaign. While stationed in smoldering Tokyo, devastated by subsequent firebombing, he heard about the U.S. Army starting construction of a hydroponics facility nearby. Hydroponics involves growing plants in sterile sand, gravel, or vermiculite and feeding them with solutions of liquid nutrients. Fifteen miles from Tokyo, Dad found the army's vast soil-less growth facility for supplying vegetables to occupation forces. U.S. Signal Corps photographs show a glass palace that stretched to the horizon in a series of peaked glass roofs; each greenhouse covered five acres.

Dad saw an opportunity. This was modern farming—the future, in fact—with all variables under control. Hydroponics could work even in cities. And thinking like a businessman, he knew that hydroponic farmers would need to buy products constantly. Upon his discharge, he returned to California and built a greenhouse at his home in La Jolla. With two friends, he started a business in San Diego selling agricultural chemicals and a hydroponics fertilizer he dubbed Nutrient Formula. He wrote and published a guidebook, *Success Without Soil: How to Grow Plants by Hydroponics*.

Hydroponic plants outyield those grown in soil by ten times, the book reports. Perfect soil is rare, and much farmland is so abused that its crops are unfit nutritionally. "It is necessary to fertilize, cultivate, irrigate, rotate, pray and perspire," Dad wrote, "in order to keep good soil fertile or to improve worn out soil." His pitch was ultimately about control. Pests, parasites, and diseases always arise when plants or animals are grown on a large scale— thus the mania for control that lies near the heart of farming. Weather is the constant farming variable, of course, and hydroponics almost eliminates it as a factor.

When I was a boy, I'd look for the green leatherette cover of *Success Without Soil*, pull it from Dad's bookshelf, and study the photos, which show chrysanthemums blooming, and lush sweet pea, cucumber, and tomato plants rising toward glass ceilings. Dad, tanned and impeccably dressed in suit and tie—even a tuxedo in one photo—fusses over irrigation valves. Other pictures document the army facility in Japan.

Soldiers longed for fresh vegetables, Dad wrote, but they were forbidden to eat local food, grown under centuries of "unsanitary and primitive fertilizing practices." This allusion to the use of human excrement was in contrast to the hydroponic plants grown in sterile greenhouses. "I wish," he added, "that those who are not yet convinced of the value of soil-less growing could see the harvests taken from those fifty-five acres of concrete."

His authorial persona—confidential, humorous, and self-deprecating (he notes in one caption that his bald scalp could benefit from a shot of

Nutrient Formula)—seemed so much more lighthearted than the solemn man I thought I knew. Dad was flooded with correspondents, he claimed in the book's second edition: of the first "so-many thousand" buyers of *Success Without Soil*, "exactly half" sat down and wrote him a letter. He printed one from a man in Maine:

> Your book and everything was fine but I raise rabbits and arthritis is giving lots of trouble. Now I don't understand why it wouldn't be a good idea to feed your Nutrient Formula and everything directly to them so that the fur would be better. Also I could save money this way. Besides, I want to go and set up my business in Florida but I don't know what effect it will have on my rabbit work but it should help the arthritis and everything. What do you think? I had a friend that moved there once but he won't write me about it. Please answer at once as I must make plans and everything.

"I know very little about Florida," Dad replied, "less about rabbits and practically nothing at all about arthritis—and *everything*!"

On a sunny day in the first week of April 1997, Kathy and I drove to Ted Foote's office in Athens to sign papers, pay money, and become legal landowners. Before we began, I asked him to order Massey to remove his cattle.

"I've told him," Ted said. "I'll call him again. Massey's used the place so long he thinks it's his." Ted's voice, which began as a powerful eruption in his mouth, seemed to dampen in his jowls and shoot partly up his nose, emerging in the powerful but muffled cadences of a lazy tuba. Later I'd hear him on the radio discussing Civil War cavalry skirmishes in our area. He was nobody's fool, and although curious about us, he didn't pry. Surely he knew exactly who we were—that is, who Kathy was.

He chuckled over the bidding war for Lost Valley. "You guys beat out a developer by only a hundred dollars," he said.

Kathy accepted my grateful look with a smile.

"He bid $80,001?" Kathy asked.

"Yes. He was upset, really angry," Ted went on. "I said, 'Hey, they won fair and square.'"

I felt we'd snatched something precious from the slashing jaws of commerce. And I felt blessed.

"We'll never sell the place for development," I volunteered, embarrassed by the sudden, self-righteous surge of emotion that thickened my voice.

Ted looked up from the papers he was shuffling. "Well," he said, "sometimes people change their minds."

PART TWO

SHEEP OVER THE MOUNTAINS

If I have sheep for forty years, my mind may wander slowly all over the place, like a sheep, but I may be worth listening to.
—Charles Allen Smart, *RFD*

CHAPTER FOUR

You may think your Jack Russell terrier and your cat are friends, but you may come home one day to find your cat dead.

—Anonymous

CLAIRE AND TOM AND I SAT AT THE PICNIC TABLE BESIDE THE CABIN, eating bagels. He had cream cheese on his cheeks and hands, and there weren't any napkins. Never mind; we were homesteaders. Real landowners anyway, and half frozen. We'd spent the night in the chilly cabin, and Kathy, who'd slept in town, had brought us breakfast. It was a glorious late-April morning, sun-spangled, drying, bathed in birdsong. I luxuriated in the sun and gratefully held a cup of coffee.

We heard the crunch of tires on gravel and saw a black Jeep pull into the farmyard below; it slowed as if to park, but then turned in front of the pond to exit. It was followed closely by another vehicle, a pickup with an orange cooler in its bed, and then by a little red car. Another vehicle entered the farmyard as the black Jeep passed it, returning to Snowden Road. We watched this tentative, circling parade and smiled and waved. Doty looked at the column and wagged her tail slowly. I wasn't sure if any of the drivers or passengers saw us up on the hill, but they stared at our van and my truck, parked together at the stairs.

The night before, I'd gotten Claire and Tom set up on foam mattresses in front of the fireplace. Having kindled a fire to roast marshmallows, suddenly I felt narcoleptic. As I fought to stay awake, Claire began regaling Tom with news from her class's sex education unit. Having myself been told about sex by two older girls in Georgia—who'd gotten the scoop, hazy as it was, from their family's maid—I was in favor of formal instruction, in theory. At least it got me out of talking to the kids about sex. And Tom's response was a third-grader's incredulous laughter. But Claire, referring to her pink spiral-bound notebook, was imparting graphic details. For the first time, I doubted her freewheeling science teacher and, incredulous myself that non-procreative acts were being described to her in detail in fifth grade, thought of halting or at least moderating her lesson. But I felt drugged, stupefied by my spring allergies. I was no match for Claire's energy as she sat with her blanket pulled around her waist and, part schoolmarm and part giggling schoolgirl—her face shining with mirth—read to Tom as I lost consciousness.

In the middle of the night, Doty's startled barking awakened me. A white flash of light swept the cabin, and I struggled out of my blankets and crawled across the floor to a window. A car was in the farmyard below, its lights off now, but its red brake lights glowing like feral eyes. I watched, my chin on the windowsill. Suddenly the car's headlights flared, stabbing across the pond, and the car turned and roared away, leaving disturbed darkness in its wake.

Now, in broad daylight, here was more perplexing traffic. "Mom," Claire asked, "what are all these *cars* doing here?"

"I think this was where people parked to sneak into Lake Snowden," Kathy said.

"Their parents parked here before them," I said, bleary, clutching my coffee. "They're like salmon, returning to where they were conceived."

As that spring of 1997 progressed, we explored the farm. A feeling of continuity arose from the aging wooden buildings, from the massive gateposts and rusty wire fences, from Mabel Vaught's old-fashioned quince shrub and her purple-blooming iris, from the shaded stony lanes and the sunny pastures. Generations of people had been at home there, molding and being molded by the land.

And one day Massey's cattle were gone. I ran into him twice more in the farmyard afterwards, both times checking his baler. "He visits that baler more than I visit my mother," I told Kathy, who said, "He probably thinks you'll try to claim it."

Walking the hills was a revelation; the land was much steeper than anything we'd known in Indiana. Ice Age glaciers, skidding like melting ice cubes from Canada, crossing the landscape in slow motion and scraping the earth, never made it here, which explained our hills and also the flat farmland to the north. Lost Valley showed me there's a lot to be said for glaciers, which would've beaten some upheavals flat. Instead, we'd gotten beauty; glacial meltwater had created gentle valleys and carved some deep canyons in the neighboring Hocking Hills.

Contractors we consulted about fixing the barn's cracked concrete-block walls, or jacking up the cabin and somehow installing a foundation, invariably said that French drains should've been installed for protection from the force of water. Having grown up on well-drained sand, I'd never heard of a French drain, which is simply a buried pipe, perforated to collect water. Folks would put them in for you, but what was the approved technique? How much gravel under the drainpipe—or is it just on top? Crushed limestone? or washed river pebbles all the way? Rigid PVC or flexible corrugated tube? In a swale? In a ditch? Suddenly I saw a need everywhere for terraces, retaining walls, steps into hillsides, berms, trickle tubes, spillways, and dry wells.

Mike Guthrie talked about land being "slippy." Sometimes, he said, water-saturated soil just couldn't cling to a slope anymore and would let go. Earth and stone buried whatever lay at the foot of the hill; slipped soil sometimes covered patches of road. We were leaning toward building a house instead of transforming the marginal cabin, so I worried about how to place a structure so the land wouldn't drop from beneath it.

Although hills are suited for grazing animals, high and dry when it rains, water really gets moving in such country. The inexorable force of gravity and the effects of Massey's cattle overgrazing and making paths had left their mark. But when I took soil samples from the pastures, I received a nice surprise: the organic matter report came in at above 6 percent, amazingly high for farmland. That much humus in the soil indicated the farm's hills had escaped being plowed for corn, and bespoke Kenneth Vaught's loving applications of horse manure. Uncultivated land was unusual even in this country, where slopes that should never host row crops had, for decades. Grazing, even with careful management, also entailed risk: this was meant to be forested land, after all; Lost Valley really wanted to be clothed in trees.

I spent a sunny Sunday afternoon in May groveling in the dirt, trying to fix a gully that threatened to form at the base of two hills in the south pasture above the cabin. Their slopes fell together in a crease that channeled runoff; the water ran into a brushy draw, which in wet weather became a rushing watercourse. The ditch was starting to eat its way uphill into the pasture. Next it would climb toward the declivity in the upper pasture where the problem originated. I pictured a raw brown gash in my green field, a vision that triggered my horror of erosion and my farmer's determination to retain every precious grain of topsoil.

A technician with the Farm Service Agency (Dad and Louis Bromfield knew it as the Soil Conservation Service) advised me to build a weir: a barrier that slows water by causing it to pool before it spills over a notch. So I nailed planks to fence posts I sank at the vulnerable junction. I would defy the hills above that accelerated water so destructively. It felt good to be engaged in the dirt, working at last on our own land again with a clear purpose. Grounded. The sun felt fine on my back. Saplings and bushes gripped the sides of the channel behind me, the branches entwined with luxuriant vines. Shy and sleek catbirds, gray with startling orange patches under their tails, crept deep in the tangled foliage and mewed. The farm's wildlife was prodigious: flocks of wild turkeys running across Snowden Road; a gawky young gray fox trotting down the slope behind the cabin; snakes swimming across the farmyard's pond; a neon indigo bunting flitting in the trees. What a farm!

On Monday, back in the office, every inch of my body sore, I was grateful to sit in front of a computer, write pitch letters, and drink coffee. It was a four-Advil morning. I read an e-mail about the university's classes in cardiopulmonary resuscitation and signed up. I imagined saving Claire and Tom with CPR after a mishap. In this region, you had to look out for yourself.

Two weeks later, my CPR instructor, Gerald, a compact, pot-bellied man who wore a bushy brown beard, began by asking everyone—it was an all-male class—to please pull up their pants and tuck in their shirts so that no one got mooned when we bent over the practice dummies. "Hey, that's not aimed at us, is it?" joked a husky maintenance worker. He and his colleagues probably were required to take the class; I was the odd man out. They presumed that *I*—bald, bookishly bespectacled, slight of build, and button-downed—wouldn't be baring my butt crack. Gerald himself soon mooned the class repeatedly as he hovered over a plastic mannequin; no mere divot, this was the full declivity.

"Don't worry if you flunk the written examination," he told us before our first coffee break. "I failed it myself the first time. You can take it until you pass." About halfway through the class, I realized that the training wouldn't stick with me until recertification in a year, assuming I took the class again, which few do. I told Gerald my concern. "Yeah," he said, "none of you will remember any of this in six weeks."

Gerald epitomized what I was beginning to think of as Appalachian Zen. This was the won't-be-rushed, live-in-the-moment, and go-with-the-flow attitude that was the bright side of such maddening practices as people stopping their cars in the middle of the street to chat, not returning phone calls, and failing to appear on time for appointments, if at all. Willy Blosser had been my first instructor. I'd read enough to know this wasn't true Zen, but it seemed to flow from the same impulse: to be aware of our existence and to savor it as we live it; to enjoy nature and to value people beyond their utility; to refuse to be owned by possessions; to dampen endless human striving with humble acceptance; to forgive life's passing trivialities.

I recalled that Bailey had shot me down when I'd tried to explain Appalachian Zen in a phone conversation. "You mean people who are incompetent?" he'd asked.

"Willy isn't incompetent."

"So first, he let his own dog kill your chickens. And second, he just doesn't care?"

"Not exactly. People here don't have much, but don't sweat the small stuff. They're not stressed out like me."

"Oh, right, I get it. They're ennobled by poverty."

Bailey's doubts usually sharpened my thinking. What did I mean? Was there a spiritual dimension to this elusive quality I sensed? Or were they just poor? I hadn't known many poor people, I realized. Now I was surrounded by them. Maybe I was being patronizing in trying to see their virtues, but not for the first time I found myself envying a more insular blue-collar world. Kathy and I had noticed back in Bloomington how the locals got to keep their children while those like us, middle-class professionals, lost theirs to colleges and then to far-flung opportunities. A clear tradeoff. Yet in America, messages seemed increasingly to flow one way: from experts and pundits and celebrities and authors and academics and politicians. Where was the wisdom of those who stayed put, those whose dream was older than the American Dream of prosperity, upward mobility, and endless consumption?

Now Gerald was talking about the first time he'd given CPR in a real emergency. He'd performed the procedure incorrectly, he admitted. A woman kept screaming at him, "You're doing it wrong!" Busy saving his victim, a man who'd been fished out of the Hocking River and was sprawled on the bank, Gerald was becoming annoyed. He looked up to tell her to go to hell and realized she was right.

It's hard to see yourself when you're in the midst of going wrong.

I began to shop for farming implements. Although I didn't yet own a tractor, somehow it seemed important to be getting ready. Farmers had lots of stuff, I knew that.

I went to see a cultipacker on a farm north of town along the Hocking River. A cultipacker is a giant rolling pin used for smoothing cultivated soil before planting—about the last thing a purist grass farmer needs. But it must've represented to me the kind of tool a real farmer owns. This one had two rollers 10-feet wide and weighed 800 pounds. I looked at it in the weeds and began to wonder what I was doing there. The farmer approached, eager to sell it for $200. This seemed high for a piece of equipment that was almost an antique. I fretted over delivery, and he asked the location of my farm. When I told him, he said a man in that area probably was going to buy a mower from him and could drop off the cultipacker. I muttered something about thinking it over and got out of there.

A few days later the cultipacker sat in the grass at the entrance to our farmstead. I couldn't believe he'd forced the sale, or that he'd known enough about my farm's location from my description. I must've mentioned Lost Valley. He had me. How could I refuse the tool or return it? And I was embarrassed. I called him and lied.

"My wife's not happy I bought it," I said, "and she's really upset that I don't even know how much delivery cost."

"Ah, Dave probably won't charge you more than $20 for dropping it off," he said. "People in Athens County won't do a man wrong. You go down in the city and it's a different story. Now, lemme tell you about women and buying farm implements. I had the same problem myself when I started out and didn't have anything. They don't understand, but you have to show them how you save money from having something."

I stood in our kitchen holding the phone, torn between amusement and outrage. I was getting marital advice from a man who was forcing me to buy his ancient tool for a premium price. He said if delivery was over twenty dollars he would make up the difference. I put a check for the cultipacker in the mail to him. One night a week later, the man who'd delivered it called me.

"I just wanted to know if everything is all right," he said.

"Yes," I said. "What do I owe you for delivery?"

"Oh, I don't know," he drawled. "Twenty dollars ought to cover it."

I asked a man who lived atop the hill where Snowden Road teed into Ridge Road to drag the cultipacker into our open-front shed near the barn; there was a bay open, Massey having finally removed his baler, which like the cattle had simply vanished one day. Fred Paine farmed various properties and owned the big cornfield that adjoined our land. He ran a construction business, too, and possessed equipment for every need. A tall man gone paunchy in middle age, Fred would arrange his fleshy bloodhound's face into a genial smile when he'd catch me speeding down Snowden Road. "Howdy, neighbor!" he'd hail me from his truck or tractor seat and offer to help in any way. But his interest seemed proprietary, his eyes appearing to assess and calculate every encounter. Of course he knew what we'd paid for Lost Valley.

After I saw that Fred had gotten the cultipacker into the shed, I drove up his hill to thank him, past two enormous lion statues, made of concrete painted black, at the base of his driveway. I offered to pay him. "Oh, no charge, buddy," he said. I sensed that Fred was setting me up for a big score, but I surely could outwit him.

While there, I noticed a fence-post pounder in his farmyard and asked him whether it was for sale. It was. Post pounders are fearsome implements that use a tractor's hydraulic system to smash a piston onto the top of a wooden fencepost; at a farm show back in Indiana I'd seen a video in which a pounder sent a husky post leaping into the earth. Visualizing such devices in action causes a person to draw back his hands, rub them together thoughtfully. I knew that building fence by driving the posts was reputed to be much faster than digging individual holes, and driven posts were said to be five times stronger than those set in dug earth.

Fred's pounder was rusty; its black rubber hoses, which lay in an intestinal tangle, were pitted and chalky. He'd lashed it to the front of one of the dented tin sheds behind his house. Fred's long face became alert as he gauged my interest. "I built all the fences you see here with it," he said, resting a hand fondly on the pounder's blistered red paint and absently brushing crumbs off his flannel shirt with the other. "It'll drive a post into dry dirt like a hammer hitting a tack."

"Would you take $500 for it?"

"No, I think I can get $600."

A new pounder would cost at least $2,000. I needed to build a lot of fence. "All right," I said, immediately feeling I'd caved, that I should've waited him out.

Fred delivered the implement to my shed, putting it beside the cultipacker. I'd soon discover that the pounder would have to be rebuilt before it would operate. The repairs would cost me more than I'd paid for it.

Yet for a bit of trouble, I rationalized, I'd get a functional pounder at a little over half of its price new. All the same, I was beginning to feel like an earthworm that had rolled into a forest pool inhabited by starving bass. Would such wily fish, cut off from the larger lake, fail to notice anything that fell into their small, clear world? No, they'd note every ripple. A plump pink worm wouldn't last long after hitting the surface and beginning its jerky descent.

In our talks with various tradesmen about what it would take to make the charming but rotting cabin livable, Kathy and I had learned to ask for a "free estimate." Otherwise, a $50 invoice for consultation was apt to arrive in our mailbox. We were leaking money like the shiny corn that flows from a rip in the corner of a feed sack. With our relative affluence enabling us to push ahead where once we would've lowered our heads and clutched our wallets, we were ripe for harvest by alert locals. Especially me in my stubborn eagerness to rush our new life and my new incarnation.

I drove to Annie and Oscar Clark's, dairy farmers who kept a pack of dogs. I wanted to give Claire one of their puppies for her eleventh birthday that May. Dealing with a puppy in our unsettled condition wasn't ideal, but these were *little* dogs, the cutest I'd ever seen, Jack Russells, and there was a pup hanging around from the latest litter. I parked beside the Clarks' bronze station wagon in front of a shed bristling with hay. Their log cabin lay out of sight below, in a shady cove.

Annie and Oscar had arrived separately in the region, a few years before Mike Guthrie, in the first wave of Athens's back-to-the-landers, she'd told me, reminiscing one afternoon after a pasture walk on their farm. Annie, strong and smiling, a strapping woman in denim or canvas overalls, was actually from the city—she'd attended high school in an affluent Chicago suburb—but fled her parents' staid middle-class world. Her vision of a farm, then, was a place with a big garden and crawling with pets—she'd always loved animals—and her reality now wasn't much different except for tending over a hundred milk cows. (She still had a horse, too, a verboten indulgence per the *Grassfarmer*.) At first she'd worked construction as a common laborer, ascending gradually to framer. After knocking together two by fours all week, she would ride her horse to a roadhouse, where she drank beer and watched men get into fistfights. And where, one night, she met Oscar. His pre-Athens past was a mystery, at least to me. Probably even then clad in his perpetual outfit of knee-high brown rubber boots, dungarees, and a ripped T-shirt, Oscar already was a landowner when they met, having bought a few scrubby acres. As soon as he'd gotten his land, he'd checked out a book from the library on log-cabin construction. Then he went to the hardware store in town, bought a chain saw, and started building a cabin from trees he cut right there on its site. He and Annie had laughed when they'd related his brash actions: when he'd set out to put a roof over his head with his own hands, he'd never run a saw or cobbled together so much as a birdhouse. His cabin wasn't perfectly square, not even close, but Oscar had never made a house payment. And soon he owned his acreage free and clear, too.

Together they had sold vegetables at Athens's farmers' market, and Annie hand-milked a cow. They heard about pastoral farming, subscribed to the *Grassfarmer*, went to grazing conferences, and bought some nearby land for dairy cows. Now they were surely among the few American subscribers to *New Zealand Dairy Exporter.*

Their dog pack picked me up at the Clarks' sturdy two-hole outhouse; yapping spotted dogs swarmed around—maybe a dozen, it was hard to tell—and Annie opened the cabin's door and grinned. A homely white terrier raced past her legs and bounded into the pack. "That's Rena, your pup's dam," Annie said, "just saying goodbye." Inside, she poured me strong coffee at a table in front of a picture window that overlooked a placid oval pond, a dock hovering upon its faintly glowing surface. She showed me our prospective pup's records on a lined card in her trim handwriting. She'd trained herself in basic veterinary techniques—vaccinations, wound stitching, midwifery—to care for her aging horse, their cows, and her dogs. Annie used to sell Dobermans, she told me, but needed a breed more suited to worrying the groundhogs that burrow under pastures and open holes where a cow might break her leg. Jack Russell pups sold well, at the same price she got for Dobermans, and the terriers didn't eat a tenth of what the Dobermans did. They gave Annie self-sustaining varmint control and some cash.

"I can't believe these cute little dogs kill anything," I said, petting the puppy on my lap. He was the size of a guinea pig, white, with brown ears and two quarter-sized tan spots, one on the top of his head and one on his rump.

"You're going to have to be careful with him," she said with a glance at the pup. "These dogs mature at twenty pounds or less, no match by themselves for a big groundhog. If mine can't surprise one outside his den, they tunnel in and take turns fighting until they wear him down and kill him. You don't want yours to get down in a hole alone and get his nose bitten off, or get his collar wrapped around a root and die of thirst before you find him."

I removed my right index finger from the puppy's mouth, his teeth like needles, and looked into his puzzled eyes, still a newborn's dull blue. Well, we needed another farm dog, a pal for the kids and an ally for Doty in warding off raccoons from the chicken house.

I got out my wallet.

"Which dog is his sire?" I asked on my way out, surrounded by her terriers' avid faces, the pack swarming the cabin's stoop; I recognized his mother.

"He's dead. The pack turned on him one day." I looked at her and must've seemed surprised—I *was* surprised. Sure dogs fight, but they aren't supposed to kill each other. Are they?

"Sometimes they do that," she said, as if explaining.

"Don't they show mercy when the loser quits fighting?"

"These dogs don't quit."

"Oh," I said, at a loss. It occurred to me I didn't really know much about dogs, at least not this kind.

As I drove back to town, I observed the unsteady pup struggling to stand on the seat. He was cute, no sign of imminent wicked transformation. Afraid Claire would give him a cutesy name, I dubbed him Jack.

At home in Athens, I tucked him into my jacket and slipped into the house's lower level, where Claire and Tom were watching television. I walked in front of Claire, stood there a beat until she looked up to protest, and pulled the puppy out and held him toward her.

"Happy birthday, Claire."

There was so much to do, so many things to try—such an energy in me—that I could work on the farm night and day, except that I couldn't. I was preparing for the press's first big exhibit, BookExpo, in Chicago, where I hoped critics and booksellers would visit our booth and place orders. With my office floor crowded with cardboard boxes packed with review copies and supplies, I filled out shipping manifests and placed calls to reviewers in New York and to our scattered sales representatives, setting up meetings.

A day after I'd left for the exhibit, Kathy received a call at her office from a tree service I'd hired to prune the farmyard's crowded trees. I wanted to preserve the cloistered beauty of the clearing, but unless I took action, some trees would be shaded out or deformed. Now the foreman of the tree crew was telling Kathy he didn't know where to begin. In the midst of her busy day as an administrator, she was irritated. She knew I'd gone over the pruning job with him. What was wrong with people here? How could there be a misunderstanding? She told him to do exactly what I'd asked, hung up, and got back to work.

Her telephone rang again. This time it was a contractor we'd been talking to about building a house at the farm; he was out there scouting sites. He told her she'd better come out, because trees were down on the ground all over the farmstead, including ones I had wanted trimmed and others I'd told them to protect.

What she couldn't know was that the previous night, Lost Valley had been in the path of a fierce, fast-moving spring thunderstorm. As the squall swept overhead, coming off Lake Snowden from the southeast, a gust had punched straight down into the farmyard and hit leaves laden with rain. At the edge of the farmyard pond, the white oak with three trunks—each ashy-barked leg about 30 inches in diameter and 60 feet long—buckled under the clout to its heavy canopy. One trunk dropped directly uphill toward the stable, landing atop a 30-foot white pine, which shattered to its base as if dynamited. Another timber, its crown glancing off a black oak in front of the barn on the opposite bank, collapsed into and across the pond. The oak's third leg crashed into the farmyard's rugged old sugar maple at the base of the cabin's hill. The maple's roots lifted from the sodden earth, and the tree fell toward the cabin. Blocked by the majestic 80-foot white pine above it, but going down like an exhausted boxer clinging to his foe, the maple stripped the pine's branches from the northeast side of its trunk, and came to rest on the face of the slope below the cabin.

The oak's overarching canopy had helped create—along with the hemlocks, other oaks, and the maple it felled—a shaded courtyard of the parking lot. This farmyard, a workaday hub from which the farm's activities radiated, had felt like a clearing in the woods, with the same shifting pattern of light throughout the day and a sense of time being suspended.

When I returned from Chicago at the end of the week and raced to Lost Valley, I was shocked by the maelstrom's destruction. The wreckage of branches lay everywhere; the sun blasted in at new angles. Nature had made its own unsentimental decision to prune. How odd that the night before it was to be trimmed, an oak that had stood for decades devastated our little farmyard. Even stranger for Appalachian Ohio, a work crew was swarming over the disaster within hours.

Unnerved and saddened, but emboldened by the destruction to act on my perceptions, I had the men return and fell an ash and a walnut that were shading out pines. The building contractor urged me to mill the logs and those from the earlier windfall into lumber onsite. I found an old man with a portable WoodMizer sawmill who said he'd tow the saw into the farmyard and make boards if I could get the trunks lined up in rows for him. Fred, our neighbor on the hilltop who'd sold me the broken post pounder, rode over on his big green tractor, and I scrambled among the fallen trees, wrapping a chain around their trunks so he could drag them into place. The gouged farmyard gravel, the naked logs, and the tree litter made the farmstead look like a logging camp when we were through.

When the sawmill operator arrived, his son helped me roll the logs toward the mill's hydraulic arms. A day of sawing wood began. The Wood-Mizer, a big band saw, transformed a log into a rectangular block with screams of its jagged blade; then the saw shaved boards from the sides of the block. I unloaded boards all morning, surprised by the weight of green oak and by its spicy fragrance: the aroma of tart apple butter rose from the damp planks.

The operator was appropriately rustic-looking, white-haired and unshaven, and dressed for his part in red suspenders—a miniature, wizened Paul Bunyan. His name was Malcolm Johnston. When we broke for lunch, the old-timer began to recount local history. Malcolm said that when the university was founded, area residents had been forced to pay an additional tax to support it. I was surprised that such history was remembered after almost two hundred years, and even more surprised by his hot anger. Malcolm's logic didn't parse, either. I knew that the university's establishment had permitted the region to be settled in the first place. That was why the institution was founded, in fact, to bring more whites to the area; Indians had been scalping settlers and skinning them alive nearby, along the Ohio River. A tax on what amounted to a fortress in the wilderness made sense.

"Those professors live up on the hills in their big houses and look down on us," Malcolm went on, his mouth twisted.

All income in the region flowed from the university! And was he blind? I was dressed like a farmer, had been working like hell, was sweat-soaked and breaded in fresh sawdust, but I was one of *them*, an outsider, a person with money. Lost Valley was beautiful, and beauty is expensive, unless you inherit it from prudent forebears.

The night the logs were milled, Kathy, Tom, and I had terrible dreams. Tom's screaming awakened Kathy. Neither of them could remember what had terrified them. In my dream, one that had haunted me in childhood in Florida, our vehicle plunged off a bridge—I was driving this time—and the car sank in the black water. I got the kids out, but the water was too deep for us to make it to the surface. Struggling, I awoke, shaken. *I already know I'm in over my head,* I thought. *Thanks for the nightmare.*

I'd just gotten the farmyard navigable again when Fred Paine appeared one Saturday morning in his black pickup with a man named Ben, who wanted to inquire about logging our place. From seeing my practical response to the windfall, Fred couldn't imagine how romantic I felt about trees. Always had. In Satellite Beach, Mom and I would drive around, scouting for Brazilian pepper saplings—actually invasive weeds, rampant in vacant lots—and would transplant them to our yard. Already I'd moved a seedling of the fallen oak to the hill in front of the cabin, and had pounded stakes beside others to mark them for protection. Lost Valley would be overhung again with oaks.

But I understood, in a way I never had, the logic of cutting some trees in their prime. Trees can be managed like a crop, selectively thinned and harvested. Logging can help a farm's income, open up additional pasture, and supply boards and timbers for barns and fences. The rustic buildings I admired on our farm were made of wood cut there. Still, I'd never consider turning loose on Lost Valley the rampant destruction of an unknown logger. I didn't tell them no immediately, wanting to preserve neighborly relations, and we chatted on the hill beside the cabin.

"I met with Mabel and her brother here last spring to talk about logging," Ben said. "She was in a wheelchair."

"I never met her," I said. "I understand that she and Kenneth really loved this place."

"You know, I bid on it," he said. "These trees right here in the farmyard are worth $10,000."

So that's his offer, I thought. It meant the trees were worth three times that. He'd planned to log Lost Valley, divide it for sale, and move on. Was he Ted Foote's angry bidder?

"We've lost a lot of trees," I said. "I'll think about it, but my wife likes these trees in the farmyard, and they do give nice shade."

Ben nodded. "I think it would've killed Mabel to see the tops of those trees down on the ground." He swept his arm to take in the 100-foot tulip poplars behind the cabin.

After sending Ben and Fred on their way with a solemn lifting of my right hand, I looked at the cabin, which was lit more fully by the sun, yet still embowered. We'd have to build a house if we wanted to live at the farm, and were dreading repeating that costly and draining experience, especially as newcomers to Appalachia.

But one night in town, getting reading for bed, I'd begun making the case to Kathy that we should appropriate the cabin's site for a house instead of taking precious pasture.

"Tear it down?" she asked.

"Yeah. It's shot. And it's in the perfect spot for a house. It commands the farmyard, where we need to have a presence. Plus it's beautiful there."

"The cabin is so charming though."

But I was thinking like a farmer.

We convoyed to the farm. I led in my truck with Doty, who gazed at me from the passenger seat. Kathy and the kids and Jack the puppy followed in the van. Summer, my favorite season, had returned, stretching out like an ocean. And the whole family had pitched in on our new land.

We parked in front of the little farmyard pond, the locus of the kids' and dogs' play. And of Kathy's work: at my urging, she was removing an old stone terrace between the gravel farmyard and the pond, because it harbored snakes. The pond teemed with life, a diverse ecosystem that the kids had discovered included horrifying leeches and exciting water snakes. At first I'd told Claire and Tom they could catch and play with the reptiles as they engaged in their primary activity—netting the pond's frogs and letting them go—but every time I asked anyone about the snakes they warned me about poisonous copperheads. I'd consulted a book, which only proved that nothing is more confusing than identifying snakes.

My verdict was that the stone wall, the snakes' apparent refuge, had to go. I shake my head now at my surety, at Kathy's good-natured assent; it must've looked to anyone—certainly it would have to Mabel Vaught, whose spirit inhabited the place—that we were intent on removing the charm that had attracted us in the first place. Yet such simple work made Kathy happy after her week of meetings and paperwork, and I had plans to make Lost Valley even more beautiful.

Kathy, wearing leather gloves and dressed in blue sweat pants, running shoes, white T-shirt, and one of my green ball caps, carried flat pieces of tan sandstone up the pond banks and dropped them along the base of the hill in front of the barn, as edging for the daylilies we had transplanted from the house in town. The kids entered the pond on rubber rafts and tried to tempt Jack into the water. He was no water dog like our late Labrador, but if he got hot enough he'd wade in and immerse his pink belly. Otherwise he spent his time stalking Doty, snapping at dragonflies, sticking his nose into holes, and digging madly after imagined gophers.

I grabbed the handlebars of my heavy-duty Stihl brush cutter from the back of my truck and headed up the hill behind the cabin to clear fence lines. Since I planned to raise sheep, good fences were the first order of business. The farm's fences were buried in grapevines, brush, saplings, and multiflora rose. I'd been surprised by the tired barbed wire that had confined Massey's cattle. Apparently the vegetation interlaced with the rusty wire had turned them.

The thorny wild rose was my worst enemy. First introduced to the United States from Japan shortly after the Civil War, as rootstock for orna-mental roses, multiflora rose spread in the 1930s with help from the USDA's Soil Conservation Service. The agency promoted the shrub for erosion con-trol and for wildlife food and cover. But *Rosa multiflora*, a species with no natural enemies in America, was too prickly to remain where it was planted. The USDA—now the plant's sworn enemy—estimated it had infested 45 million acres in the East, South, and Midwest.

Lowering the face-protecting mesh screen on my orange hardhat, I waded into the treeline along the edge of the pasture and cranked my cutter to life. First, the roses' protective and confusing outer growth had to be cut away from entanglements with trees, weeds, grapevines, and barbed wire. I marched inward, pruning limber green rose stems that arched into the pasture from mother plants and could root to form new colonies where they touched the ground. As these fell away, I cut deeper, into the shrub's base, a mass of thorny shoots that had thickened and become woody with age. Then I attacked the center, a stiff fortress of durable canes, shaded bare by the plant's own rampant growth.

As vegetation closed around me, I struggled to find the source of the spiny tendrils that climbed high into trees. The same species that grew into a thick bush in our fields' sunny borders also adapted to deeper shade, where the plant became vinelike. The weight of the vines pulled tree branches downward. I staggered, clutching my handlebars, the brush cutter in front of me a blind man's cane as I slashed toward the source of a plant.

The cutter revved and smoked. A thorn pierced my canvas pants and drew blood; the back of my neck itched. My eyeglasses fogged, and sweat ran in rivulets from my armpits. I knew I'd found the fence when I saw sparks flying from the weed whacker's head, a three-bladed steel saw that looked like a propeller. I took a step back and swung the tool in mighty arcs over my head at the draping confusion of multiflora rose, tree branches, and grapevines.

A thick steel cable appeared before my eyes; it was a silver multistrand cable, the sort that supports telephone poles. Like the one that stood behind me at the edge of the pasture and fed a power line to the cabin below. Then the cable disappeared into the jungle from which it had sprung. I'd felt no impact, and experienced a moment of adolescent denial before the limp cord slithered heavily across my shoulders.

In the wooded slope behind the cabin, smoke rose and sparks flew as the electric line sagged crackling into the trees.

From the cabin I called the power company in Columbus and listened to the limp power line sparking as I explained how to find our farm. Then I sat at our helm, the picnic table, pleasantly tired. Kathy, her obstinate stone-work done for now, had fetched us lunch from Albany. The village's Dairy Queen was having a banner year because of our arrival.

The kids left the pond and shifted their water play to a garden hose. Tom was prone to sunburn, but wore only his baggy teal swim trunks as he squirted down the concrete slab around us; Claire, already tanned deep brown, was watching the water flow between her toes and giving him urgent directions. Doty was running from Jack in great teasing circles through the shady yard, her orange coat a blur. He crouched, calculating how to intercept her; still too small and too slow, he took his shot and leapt, but missed, and stood looking baffled and vengeful.

Kathy and I attacked our burgers, while Claire and Tom pulled from their "Chicken Fingers Baskets" a carbohydrate feast: fried strips of meat,

French fries, cups of gravy they ignored, and slices of toast they saved for the dogs. Doty and even Jack plopped down, too tired to show interest in our meal. The afternoon hung in the sky. There was a welcome languor in humidity's return. My emotion surged, a leap in my chest, at summer's dramatic possibility.

"Your stones make a nice edging in front of the daylilies," I said to Kathy.

She nodded, chewing. "After lunch we need to water them," she said. "Claire and Tom, you can help with your buckets." Then to me: "They didn't bloom much."

"Next year," I said. It was really too shady for daylilies under the oaks, though I pictured the dry bank above the pond massed with their yellow and orange blossoms.

"Look at Doty!" I cried.

Back in Indiana, Doty had been a grinning dog; after a good day she'd lie on her back in the family room and bare her teeth, a disconcerting sight until we understood its meaning: pure contentment. As sweet and sensitive as Jack was scrappy and stubborn, Doty hadn't acted happy since our move. But now, belly up, sprawled open to the universe, her white front paws tucked on her white chest, her back legs splayed in a frog kick, Doty grinned.

CHAPTER FIVE

You can clear multiflora roses by hand, but not in any quantity.
You can pluck and eviscerate a chicken with your teeth, too, but I
wouldn't advise it.

—e-mail to Bailey

WEARY OF MY ENDLESS WORK CLEARING FENCE LINES OF MULTIFLORA
rose, I drove my truck to a pasture walk on a warm Friday morning that
August. The farm, west of town, recently had been purchased by a young
couple planning to start a dairy. About eight of us showed up to give tips on
fencing and opinions on the state of their pastures. As a neophyte myself, I
didn't have any advice to offer, but hoped instead to learn.

The host couple had moved down from Columbus, where they'd studied
animal husbandry at Ohio State and then worked and saved to buy land.
The taciturn young man, in his late twenties or early thirties, rail-thin and
wearing a bushy blond beard, exuded determination. He'd grown up on a
farm in the region, he said. His shy brown-haired wife, in a yellow sundress,
said she'd stay at the house during the tour to watch over her new baby.

Before walking the premises, everyone introduced themselves. I received
polite smiles and nods, but today felt uncertain among these acquaintances
and strangers, most of them wearing knee-high rubber boots, ugly but

indestructible farm footwear that I'd grown fond of in Indiana. Having left my own boots in town in deference to the heat, I felt self-conscious in my grass-stained white-leather Adidas sneakers.

I owned a grass farm at last, but still no livestock, and had just passed up the chance to buy sheep. One evening the week before, an elderly woman, Zinnia Rhodes, had called me about her flock. Her husband was dying of cancer, she said, and she needed to sell her ewes. I'd seen her advertisement for a ram in *Farm and Dairy* soon after our arrival, and had talked with her. Even though we now owned a farm, I wasn't much readier for my own flock. But my mind raced when she called and offered me a package price.

"Richard, you're moving so fast," Kathy had said. *Too fast* is what she meant. I was spending more time at the farm alone. One evening as I was racing around, shoving tools into my truck for a quick trip, she arrived, and as she walked into the house said something uncharacteristically sharp: "I hope you're enjoying your new life." And I'd snapped back: "I'm trying to build *our* new life because *you* moved us for *your* career!"

She apologized later, blamed a hard day. I apologized, too; I knew she was right. I couldn't seem to stop myself, though, to quiet my sense of urgency.

Annie and Oscar Clark, our puppy's breeders, who tucked their identical faded tan Carhartt work pants into matching pairs of brown rubber boots, were the grazing group's stars, and prominent this day, with Mike Guthrie away at a building site. Oscar told the pasture walkers that they were now milking 125 cows in their "grass dairy."

"We're livin' the dream, and we're making it work!" sang Oscar, compact but muscular, bald but bare-headed. At each pasture walk, he gave impromptu lectures about everything from grass species to grazing economics. There were rumors that he and Annie might be written up in the *Stockman Grassfarmer*—the grazing-world equivalent of a rock band getting featured in *Rolling Stone*. However obscure, the *Grassfarmer* was our common touchstone, egging us on with stories about the low costs and high returns when livestock roamed sun-washed fields.

I couldn't help but remember the pasture walk the previous fall at Mike Guthrie's. Mike, who years ago had been featured in the *Grassfarmer*, had stood in a field and said, "Look at the color and quantity of this forage, without any purchased nitrogen. That's from clover."

"Purchased nitrogen can pay," interrupted Oscar. "You can take just a piece of the agribiz answer. You don't have to play the whole game."

"The game I play is *not* spending money," Mike said. "I'm good at raising cheap sheep."

"You've got to spend money to make it," Oscar said. "Debt's a tool—"

"Free solar power is *my* tool," Mike said. "You've got cash flow, sure. And debt up to your eyeballs. One problem and you've lost your whole farm."

"A commitment involves risk." As Oscar spoke, out of the corner of my eye I saw Annie amble off, talking to another woman. The rest of us stared at the ex-hippies butting heads.

"I do so little that I'm paid twenty dollars an hour," Mike said, a wicked grin growing. "I'll bet *you* don't make that, and I've got practically nothing in infrastructure."

"That's apples and oranges. We make our living farming, it's not a part-time—"

"Who *wants* to be a full-time farmer?"

As Oscar looked stunned, the county agent cut in. "Okay, let's go see Mike's new pasture of Tekapo orchardgrass."

Big grazing dairies remained the *Grassfarmer*'s latest get-rich-off-grass obsession. Instead of standing around eating grain and hay in muddy feedlots, cows were sent to forage for themselves. The dairyman, once reduced to toting feed and scraping manure off concrete, spent his time nurturing pasture salads of succulent clovers and tender leafy grasses.

Now, as we began climbing a hill that overlooked the farm, I was keenly aware that I'd chosen a safer path, part-time, and seemingly lesser, despite Mike's endorsement. And what I *was* doing felt overwhelming. The fields and deep blue sky exuded summer's timeless languor. I could be home with Claire and Tom, laughing at videos in the cool basement. *This is an adventure*, I reminded myself.

Annie fell in beside me, her boots swishing through the ankle-deep grass.

"Lots of feed for the cows," I said.

She glanced at me. "This forage is *way* overmature," she said loudly.

"Poor feed value. And Kentucky 31 fescue is *serrated*. At this growth stage it's like eating sawgrass." She bent and tore off a blade. "Chew on this," she ordered. I tasted its greenish tang and felt its sharp edge.

We stood looking down into the tangled sward. Annie was about my height, five-foot-nine, but more robust. I could see a dark tattoo on the rounded muscle of her upper left arm, half obscured by the sleeve of her white T-shirt, whose red printing across her chest asked, "Got Milk?" She squatted suddenly, plunged her hands into the pasture, and parted the grass, revealing soil between the clumps. "Look here," she said. "This fescue looks thick but it's just tall. A dairy-quality pasture is *dense*. That's what you need for lambs, too. This needs legumes. Protein. That's going to be hard without more cows. The fescue will swamp the clover." Annie fished around in the grass, found a thin light green shoot, plucked it, and thrust it toward me.

"Bluegrass," she said. "See the boat-shaped tip? Taste it."

"It's tender—and sweet!"

"Yes. More bluegrass will come in, and there's room here for clover, but not unless they put some pressure on this fescue. It'll just shade out everything else."

"I've been mowing our place, trying to favor the clover and bluegrass," I said. Actually, since I didn't have a tractor yet, I'd had to hire Willy Blosser, my disastrous one-armed chicken landlord, to shred our ten acres of pasture with his Bush Hog cutter. Without constant grazing by Massey's cattle, Lost Valley had erupted in coarse fescue; no more drifts of white clover.

Willy was cheaper than other men who cut pastures, but he was high-maintenance. Taking pride in cutting every scrap of grass and weedy corner, he slammed his mower into Lost Valley's stumps and surfed it over logs; it rasped, bled oil, smoked, and quit. I jollied him along through breakdowns and his unexpected need for food, which I fetched from the Dairy Queen to keep him in the tractor's seat. I even brought him diesel fuel when—surprise!—he ran his tank dry. Willy, covered with chaff, just grinned into my tense face and shook his head at life's imponderables.

When Zinnia called me about her ewes, I was haggling over a tractor, a dealer's green and yellow slightly used John Deere that I would soon buy for $16,000. I'd sold my tractor in Indiana for $12,000, so the purchase wasn't a big stretch. But I was overwhelmed by the need for infrastructure—mostly fences and a corral—and worried about how I would water sheep. Massey's cattle had drunk from the pond in the north pasture after he'd breached the pond on the south side—but even if I wanted to let sheep drink such fetid water, I couldn't: I'd be moving them all over the farm, not letting them wander at will back over grazed pastures.

When I passed on Zinnia's offer, her tone implied that I'd blown my once-in-a-lifetime opportunity to get ewes she'd never consider selling if her husband weren't dying. As with finding land, such farm opportunities seemed never to appear and then suddenly did, usually because of someone else's misfortune. The editor of the *Stockman Grassfarmer* liked to point out that the four D's created change in farming: Death, Divorce, Disease, and Disaffection.

Annie rose from the grass and appraised me. "What did you say you do for a living?" Her gray eyes regarded me from a long, sunburned face framed by glossy black hair pulled into a ponytail.

"I work at the university press. My job is to get our books reviewed. I work ahead, months before publication, mailing letters and galleys. I go to New York twice a year and meet with reviewers."

As I spoke, she stared at the ground, boots planted, brow furrowed.

"Oh!" she said, meeting my eyes, grinning like a kid who's just solved a riddle. "You're a *publicist!*"

"Yes." My face flushed under my blue baseball cap. "Publicist" was my job title, in fact. Why couldn't I just say that? Out here under the sun it sounded tawdry, somehow, even though my work involved books, not Hollywood starlets. To persuade the *New York Times* to review a book published by a small Midwestern press took effort and skill. My embarrassment shamed me further. Who the hell was I?

No grass farmer yet, for sure—unless eating two pieces of grass counted.

I told Annie I was going to skip the milk-house tour and head home. "You won't miss much," she said, stopping at the brow of the hill, her hands on her hips. Dangling from a back pocket of her canvas work pants, a pais- ley purple bandanna swirled in the breeze. We looked down at the farmstead where a low rectangle of gray concrete blocks jutted from the side of an unpainted two-story wooden barn.

"*Man,*" she said, shaking her head. "The herd's barely started, the milking parlor's not equipped, but look at that house! A startup farm can't pay for a house like that."

The couple had bought the place from a part-time beef farmer, whose day job was as a professor at the university. His house, clad in attractively rustic boards stained a mellow amber, was modest by suburban standards. But I saw the place for an instant through Annie's eyes: a young couple seduced by an irrelevant house, their larger dream compromised, if not doomed from the start. Yet how natural, I thought, for a young wife with a new baby to want a cozy nest.

"It's like that new extended-cab truck you drive," Annie went on. "A farm can't support a vehicle like that. And who really needs a pickup truck? People drive around all week with the back empty. On Saturday they throw in a few feed sacks that could fit in a Toyota's trunk or be delivered for a small fee. Do the math."

"My truck's three years old and it's our *family* vehicle too," I blurted, my face again hot. "We have two kids. And I keep vehicles a long time. I drove my last truck eleven years, until the body fell apart. I'll keep this one at least fifteen."

Annie nodded, her calm gray eyes on me. "Sounds like you've thought it through," she said.

Below us in the farm's makeshift parking lot, her and Oscar's bronze station wagon glinted in the sun near my truck. In the era of the soccer-parent van, they'd gotten the creampuff old Vista Cruiser cheap. With their homemade cabin and outhouse, they were living, not just espousing, the *Grassfarmer*'s tight-fisted philosophy, atop a dash of the *Mother Earth News*.

Leading his acolytes down the slope, Oscar, a wild gray fringe circling his browned pate, waved his arms as he lectured. I heard him say, "Farming's the last stage in the counterculture revolution!"

Driving home in a funk, I recalled a song Dad loved: "Nobody loves me. Everybody hates me. I'm going out and eat worms." *I'm going out and eat fescue*, I thought and smirked. Then: *I'm going out and whack multiflora rose!* I'd get the dogs and my weed whacker and make constructive use of my angst.

I remembered a photograph in Louis Bromfield's *Out of the Earth* that shows a flock of sheep confined in their pasture at Malabar Farm by a thick wall of rose canes. Bromfield had been an avid promoter of multiflora rose as a "living fence." To tend the hedge, he employed a special mower with a hydraulic arm that lifted the cutting deck away from his tractor and thrust it out to shred the stems. Even with the aid of such a fearsome device—too pricey and specialized for most farmers—living fence maintenance was a dubious idea in America, where farm labor had dwindled for over one hundred years, and land even then changed hands too often to count on descendants tending European-style hedgerows of rampant Asian rose bushes (each one capable of producing one million seeds a year).

Some people, forgetting the government's role in boosting the tough species for erosion control and wildlife, still blame Bromfield alone for the plant's invasion. Indeed, having read Bromfield as a boy, I surely would've planted the demon at Lost Valley if doing so by then hadn't proved to be an obvious madness.

Rereading Bromfield's books in Athens it occurred to me, without diminishing my enjoyment, that he hadn't let facts impede a good story—and he laid it on as thick as good manure. In *Malabar Farm* he brags about making topsoil: "It is largely a fact that in a period of eight years we have produced good productive soil directly from bare subsoil." *Largely* a fact? So seductive was his prose and his vision that I'd been surprised when one of our authors told me that "My Ninety Acres," the heart of Bromfield's memoir *Pleasant Valley*, is actually a short story.

The main character is an elderly farmer named Walter Oakes. Although mocked in his youth for his endless talking about his haven, by old age Walter has become a local touchstone, and the farm an attraction. It's no sterile showplace, but bursting with life, the thick fencerows sheltering quail and songbirds that eat insects that otherwise would destroy Walter's vigorous corn. The overgrown flower garden is a clash of colors, while the vegetable plot is orderly and weed-free. The creek banks and woodlot grow rank, the cattle having been fenced out; white-faced Herefords stand knee-deep in alfalfa in their own pasture. Walter can be spied wandering about the fields or on his hands and knees, examining earthworm activity.

"There was, indeed, a certain shagginess about it," writes Bromfield, "a certain wild and beautiful look with that kind of ordered romantic beauty which was achieved by the landscape artists of the eighteenth century who fell under the influence of Jean Jacques Rousseau's romantic ideas regarding Nature."

Walter's wife, Nellie, had been the prettiest girl in the valley, with her pick of heirs to vast acreages or men with business prospects. But she chose Walter, owner of a small hill farm. In time, the other farms deteriorate, and the businessmen become drunkards, but My Ninety Acres brings in bumper crops. And Walter—apparently an idealized portrait of Bromfield's father—has supposedly conveyed his wisdom to the writer over the years on pasture walks.

Reading Bromfield's books as a boy, I'd been puzzled as his pages grew increasingly pensive, especially the haunting final chapter of his last book, *From My Experience*, "The White Room," about his yearning for spiritual peace. The White Room was in an old house on a Brazilian farm, *Malabar do Brasil*, managed by his daughter Ellen and her husband, Carson. Bromfield was tired, ill from the undiagnosed bone cancer that would kill him the following year, but in Brazil he finally found relief from "the fierce activity, the driving turmoil" of life at his fabled Malabar Farm. "What was this force that drove me?" he wondered.

He read Albert Schweitzer's *Out of My Life and Thought* on this trip, and Schweitzer's phrase "reverence for life," the philosopher-physician's spiritual response to the life force, struck Bromfield as the answer to his own life and to farming's larger purpose. "Every good farmer practices, even though he may not understand clearly, the principle of the Reverence for Life, and in this he is among the most fortunate of men, for he lives close enough to Life to hear the very pulsations of the heart, which are concealed from those whose lives are concentrated upon the unbalanced shabbiness of the completely material," concludes Bromfield and the book.

The farmer as one in touch with the Life Force appealed to me, but as a boy I couldn't understand why Bromfield felt the way he did, openly ashamed of his work as a popular author, bewildered and depressed after a lifetime of accolades and the owner of a legendary farm. I didn't know then that after his promising early fiction, Bromfield had alienated the literati with his potboilers and his exhortations about U.S. foreign policy. Or that his live-in agent and editor, a countervailing force to Bromfield's farming obsession, had died. As had his wife, an invalid, who years before had retreated from his chaotic household and his occasional mocking of her helplessness. Or that his overbearing ego had alienated his daughters and their husbands. In his final years Bromfield was lonely and alone, rattling around in the silent Big House.

For the first time I was reading Ellen Bromfield Geld's wistful memoir of her father, *The Heritage*, because the press's editors had decided to return it to print. In her account, Bromfield's father, a small-town bank employee and politician, was a dreamer who lost money attempting to restore abused, bankrupt farms. Bromfield was spurred by his forceful, ambitious mother toward fame, yet in the end the driven son resurrected his impractical father's dream. Ellen Geld wrote of Bromfield's father, her beloved grandfather, in *The Heritage*: "He was the first of us to have been possessed with a mania which has continued on through his seed in the generations that followed: the uncontrollable desire to bring back to life from the sadness of abandonment old farms which had once known the great happiness of fertility."

As always when I read anything concerning Bromfield, I thought of my father. I pictured him after his father's suicide and its social shame, when he and his mother abandoned their Detroit mansion, bought from car maker Henry Ford, and took refuge at the family's professionally managed farm west of the city in what one day would become the upscale enclave of Bloomfield Hills. Though Dad was a full-time boarding-school student by then, the farm was now officially his home. Given the run of that gracious farm, the boy tried to heal. He roamed and hunted, and once nursed an injured Cooper's hawk back to health. While still a teenager, he designed, dug, and dammed a pond with the farm's bulldozer.

I had to wonder now, had we stayed on the Georgia farm, wouldn't my driven father have driven away me, as Bromfield had his own offspring? Farmers are notorious for overworking and overly dominating their sons. I could have accepted that—all of it, I believed—if he would have put his arm around my shoulders. I'd always yearned for what he couldn't give after his traumatic childhood: his loving presence. And it shocked me to realize so late that growing up on Stage Road Ranch might have inflicted a worse trauma than its loss.

Now I felt myself a child of two fathers. Proud of Dad for writing *Success Without Soil,* I had reservations about its message. I could see why my father—as a poorly nurtured child, as an adolescent sent to boarding school, as a fourteen-year-old devastated by his father's bloody death—would think that completely artificial means (glass greenhouses and chemical fertilizers) could replace the natural world.

I believed with Bromfield, however, that there's no such thing as success without soil.

Yet the two men were much alike. Only a romantic would've gone into ranching as Dad did instead of investing his once-considerable assets. Bromfield, while chasing a utopian vision of agriculture, was more pragmatic than his volcanic prose suggests; other farmers made money from his ideas. Both men succeeded and both men failed. Had they met, I think they would have taken each other's measure and understood.

One thing was certain. Nobody could fail to be affected who, at an impressionable age, ripped from his family's own mythic land, had devoured Bromfield's sermons on the holiness of humus. A moment's thought would've told me why I'd been so sensitive to mud and erosion at Diana's dairy farm, and why I'd been so determined to protect Lost Valley's topsoil from Massey's cattle.

And our old hill farm possessed the shaggy beauty of that fictional farm, *My Ninety Acres,* with which, so long ago, I'd fallen in love.

One weekend in early September I removed the log cabin's wide-plank doors and stored them in the barn's loft. Kathy unscrewed the light fixtures ornamented with horseshoes. I had finally convinced her to let me tear down the cabin and build our house on that spot. That way, we wouldn't lose any pasture. As it was, I could responsibly graze only about 30 ewes, a pittance compared with even Mike Guthrie's sideline of 130.

For the demolition, I ruled out hiring a bulldozer as too destructive for the woodsy site and instead called Frank Willis. The next week he arrived and limped around the cabin, looking up at its steeply pitched roof and running his fingers through his thick brown hair. He was disabled from a mining accident, he told me. "We can get this building down in three weeks," he said.

He was as good as his word, for the process took twenty-one days—about twenty days longer than a bulldozer would have. Frank's crew materialized from the hills. Although I was paying him, his workers, it turned out, were working for a share of the building's materials. I was glad the cabin would be reincarnated as bedrooms, porches, and chicken coops in the region. But it was a long three weeks, because the cabin was a sentimental landmark that I was causing to disappear rather slowly.

People came by foot, vehicle, and horseback to watch. Their inquiries were beyond circumspect—were unspoken, in fact, but implied. Why were we destroying a cute little cabin, perfect for weekends and as a deer camp in hunting season? My decision pained me as I watched the cabin go; it had been an asset and—Kathy was right—so charming. A place to camp with the kids. "We're going to build a house here," I told Frank defensively. He nodded, his eyes elsewhere: my plans weren't his interest.

Like Malcolm, my buddy with the portable sawmill, Frank felt compelled to air his own grievances. His beef dealt with the employment practices of the university, where he'd gone to work after being hurt in the mines. "They ran short and fired a bunch of us," Frank said. "The workers always pay the price. They need us to keep the place going, but think they can get by for a time, lay us off, save some money, then rehire. Like we don't have our own bills to pay."

Knee-deep in rubble and wads of pink insulation, my shoulders sagged as Frank went on and on and I turned my sweaty face to stare into his. He even brought up the resentment, two hundred years old, that the university's first president had barred settlers from grazing livestock on the college green. I hadn't any answers for him, and in fact was developing deep reservations myself about the university. Not it per se, but the fact that it anchored the local economy. The region hadn't ever supported much industry, any kind of manufacturing base. Wealth had neither accumulated nor been spent locally. The towns, poor, ugly, charmless, showed this.

No one institution could provide everything the area needed. And Athens seemed to lack an old guard, people who understand and love a place and have a stake in its future. I found myself yearning for an enlightened business class that would generate amenities for academic careerists and clueless newcomers like me. In Athens, generations of merchants had made their real money by harvesting dollars from college students. Bars were big business. Even more lucrative was what seemed to be everyone's sideline: renting, at exorbitant rates, tired old houses, chopped into bedrooms, to students eager to get out of the dormitories.

Ohio University drove the economy in one way or another, but a university seemed especially unfit to be alone at the top of the socioeconomic heap, given its tax-exempt status and its transient students and instructors. Professors who did put down roots, along with the counterculture outsiders who'd made their way to Appalachian Ohio, mirrored, in their own way, the contrary locals: to remain here, they'd taken a left turn at the American Dream. They bitterly opposed plans for a Wal-Mart, which the populace devoutly favored. Professors had reliable vehicles to carry them north to Lancaster or east to Parkersburg, West Virginia, to shop. The tacky trailers and broken cars that dotted the countryside were ugly, but who went out in the countryside anyway? Or so I imagined their thinking in my anxious state. At that point I actually knew more local people than I did professors, but I'd met a couple of academics, originally from urban areas, who expressed surprise that Kathy and I were planning to move outside the city limits. They seemed afraid of the countryside and its inhabitants.

Athens was a magnet offering, in comparison to the region, an oasis of culture. Yet I still compared the town, its merchants and its tradesmen, unfavorably to our old hometown in Indiana. "Here," I e-mailed my friend Bailey as Frank's crew reduced the cabin into rubble, "is a small pie that not many share, no competition, high prices, and rampant incompetence."

Finally he let me have it: "I haven't called you on this before, but you've been wallowing in negativity. I learned that lesson myself when I used to bitch at my last job, and our copy desk chief said to me one day, 'Try raising three kids by yourself on a secretary's salary after your husband has dumped you and you're going through chemo.' I needed to hear that and grow up, and you need to hear this. Yes, you had it made in Indiana but you left. That is now in the past, don't you see? You probably didn't appreciate Blooming-ton enough when you were there. You're causing your own unhappiness. Try some resignation and acceptance."

What could I do but thank him?

Still, as the cabin vanished, as dented pickups and ancient flatbed farm trucks groaned away from Lost Valley under petrified logs, splintered boards, and tangled plumbing, I cringed, imagining the news spreading about how my vision was taking shape: *Some fool is tearing down the Vaughts' cabin.* More rubberneckers arrived. Kathy and the kids avoided the spectacle, but I couldn't, having caused it.

After Frank and his crew raked up the last debris and left, I hired Fred Paine to bulldoze a curving driveway from the cabin's site out to Snowden Road through the woodlot. Now the farmyard was ready for our new home, if we could decide on plans, which seemed to arrive daily at our mailbox in town from mail-order design companies. After I wrote Fred a $100 check and handed it up to him in his yellow machine, I brought up something I'd been pondering: his cornfield along Ridge and Snowden roads at the entrance to Lost Valley. I craved more pasture and also feared he'd sell it for dwellings, or at least liquidate its road frontage, as he'd done with the other side of Snowden Road. The *Grassfarmer* had once printed an old saying about farmers that captured my land hunger: "A farmer doesn't want to own all the land in creation, just all the land that borders his farm."

"If you ever think about selling it," I said, "I'd be interested."

He nodded, poker-faced, sitting in his dozer seat. I added, to clarify this was about farming, "I'd like a little more grazing land."

"Okay, buddy," he said, in his smooth, deep voice. Easing his dozer into gear, he clanked down Snowden Road toward his hilltop.

I stood savoring our place. No vehicles except mine, no workers whose children played in the dirt, no idle gawkers. Still, I couldn't bear to glance at the unsettling void atop the hill where the cabin had stood. The farmyard felt empty without its silent, mossy presence. At least the baby white oak I'd transplanted near the base of the stairs—stairs that now led to a trampled rectangle of mud—had grown. The first landscaping for our new house.

On the Friday after Thanksgiving, farm-sitting again at Mike's place between Athens and Lost Valley, I drove into the nearby crossroads of Shade to buy dog food for his spaniel. The Lodi General Store and Community Post Office was a narrow concrete-block structure with a stoop. Inside, three men and a woman sitting at a table stared at me. Granted, I was a stranger. But I was dressed in canvas coveralls and hadn't opened my mouth to reveal that I was an outsider.

Self-consciously I looked for dog food but couldn't find it. Mike had said I could get it here. Finally I asked the people at the table where it was. "It's in back," the biggest one said. I finally found sacks in an ell tacked onto the store's rear. I paid the proprietor in a completely silent transaction. Such a lack of graciousness in Appalachian Ohio still bugged me. In Georgia, the people in the store would've asked me how I was doing, commented on the weather, passed the time. In Indiana, someone would've offered help.

I'd learned in my day job at the press, where we published books in Appalachian studies, that the discipline, which claims dominion over the region's image, dismisses a literary classic about Appalachia, *Night Comes to the Cumberlands*, by Harry M. Caudill, because it feeds hillbilly stereotypes. I went to the library and checked it out. Perversely, academia's disapproval increased my interest in the book, which was credited with making Appalachia a focus of America's "War on Poverty."

Caudill's 1962 treatise rises to Faulknerian grandeur as it traces the roots of the independent mountain folk, many descended from indentured servants who came to America from England, Scotland, and Ireland. The strongest and fiercest among them escaped slavery by hiding in Appalachia's cool, forested hills, where they founded a woods culture—a self-sufficient hunter-gatherer life—and carved out hardscrabble farms. Caudill ponders the effect on them and their land of generations of extracted wealth in the form of lumber and coal, of their relations with Native Americans. Despite his larger indictment of America, his account has become completely politically incorrect. He writes:

> Consider . . . these forces in synopsis: The illiterate son of illiterate ancestors, cast loose in an immense wilderness without basic mechanical or agricultural skills, without the refining, comforting and disciplining influence of an organized religious order, in a vast land wholly unrestrained by social organization or effective laws, compelled to acquire skills quickly in order to survive, with a Stone Age savage as his principal teacher. From these forces emerged the mountaineer as he is to an astonishing degree even to this day.

What attracted me more than the darkness of Caudill's portrait was its romance, surely drawn as much from poetry as from history. I'd still defend romance, the ancient and mollifying truth of poetry, as a better way of understanding a person, if not a people, than the reductive social-science impulses of our age. Willy, for instance, couldn't be grasped numerically, and no facts could explain his good nature, which defied what could be known about his region, his family, his physical plight. Willy, who'd viewed his dog's slaughter of my chickens with a maddening "stuff happens" grin; Willy—the crazed tractor driver, bouncing his rig off trees and crashing it over deadfalls—was the sweetest guy. Despite chronic pain, a suffering anyone could see, he grinned tolerantly at his own folly and at life's absurdity. At America's dominant money-grubbing culture that had no use for him. At my own anxious striving.

With or without my approval, the locals were capable of taking care of themselves. They'd survived for several hundred years in a region bypassed and abused by the American Dream. They were as wild at heart as *Rosa multiflora*, another prickly escapee from the confines of an established order.

The snow of our second Appalachian winter sifted through the trees and dusted the hills. I was driving to the farm that Sunday afternoon to prepare for my flock and maybe to do some pruning. Winter chores lack the urgency of work in spring or summer.

I got out of my truck in the hushed farmyard, and a deep peace descended from the hills that sheltered the clearing at the heart of the farm. The dogs bounced in the cab, sweet Doty and our scrappy child of Appalachia, Jack, who'd grown into a true terrier, all muscle and bone. As I walked to the passenger side to free them, I realized I'd forgotten my weed whacker, unthinkable last season. At the house in town, my mind circled farm projects like an oak leaf caught in an eddy against the barn. But this was winter. Standing in the farmyard under the branches of the ancient white oak tree on the north bank, I couldn't quite remember what needed to be done.

I began to wander and look, purposeless for once, lost in the dream. I climbed the long wooden steps to the gap where we'd removed the cabin. The rye I planted had sprouted and would hold the disturbed soil until we could begin building in the spring. On the lawn, the light gray bark of Mabel's saucer magnolia tree caught my eye; its puzzle of crossed and crowded branches needed attention. I decided to leave the pruning for another day, a chore to savor.

The dogs rushed together from the woods beside the yard, struggling with the forelegs of a deer; the bones, too unwieldy for one dog to drag alone, yoked Doty and Jack parallel, and they jerked their heads for control. They'd found the remains weeks earlier and periodically rediscovered and wrestled over the bony treasure. The farm was silent as a painting. In the farmyard the narrow pond was frozen a silver-gray. Just six months before, in spring's changeable glory, Claire had caught in it a big leopard frog with olive skin and shining black spots.

I found myself in the north pasture, walking up Massey's tractor trail through dry, ungrazed grass. My mother had struggled along the same route with me during a visit in early spring. Despite pain from an injured knee, she'd climbed the slope, holding her black leather coat closed against the chill. I paced off the distance up the gentle rise, calculating where water troughs should go to serve grazing paddocks. I tried to spot natural breaks where subdivision fences should run, along the keylines, creases that are important transitions where steep hills shift into gentler slopes and where those slopes become flatland.

I remembered my purpose and walked back down to the barn and grabbed three 16-foot welded-rod fence panels and dragged them uphill. Arriving at the plateau in front of Massey's old cowshed, I caught my breath. I watched a red-tailed hawk soar across the end of the field, over the grove of white and green mottled sycamores Kathy loved. The bird uttered a piercing cry as it glided toward Lake Snowden. From my elevation I could see down into the woodsy pasture below the farmyard pond; light snow blew in from the east, filtering through the trees. The whitened slopes of the south pasture rose in the distance.

Massey built this shed years ago. Its low interior had always repulsed me—a dark, fouled cave. The frozen ruts in front of it, from Massey's hay deliveries, were painful to traverse. When it was warm, you could sink past your ankles into muck; by late summer, red-stemmed pigweed, pokeweed, and giant ragweed grew tropically luxuriant there, as tall as the building's eaves.

The shed was clad in unpainted boards, a thirsty silver with blackened streaks, and its dented, undulating tin roof was supported by the gnarled trunks of black locust trees that Massey had hacked from the pastures. He'd tucked the shed into the lee of a hill that rose protectively behind it—almost sixty feet straight up—to the west. The hill curved around the shed to the south as it fell gently away. I saw how Massey had cunningly angled the building so that its open front pointed southeast, thereby protecting the interior in summer from the prevailing southwesterly winds, and in winter from gales arriving from the west, north, and east. The shed's face admitted precious winter sunlight but wouldn't fill with snow. Nary a snowflake blew inside today.

At the foot of the hill on the shed's southwest corner, a walnut tree, its downward-looping black limbs almost touching the roof, provided shade in summer from afternoon sun. The roof held years of blackened leaves, twigs, and walnuts, and sagged in its middle like an apron. The tree was doing its best to fertilize the shed back into the earth, yet the patched roof testified to Massey's determination to prevail.

I leaned my fence panels against the shed and removed my leather gloves. Inside I pulled pieces of string from piles of Massey's old hay twine. I would tie the fence panels to the locust trunks to make a sheep pen. I'd stashed other sections in the shed, and I bent some of them into gentle curves, forming a chute for ear-tagging and deworming our future flock. Having used old materials lying around the farm, I felt frugal and farmer-like—like Mike Guthrie, like Massey—for having avoided purchasing anything for the shed's latest incarnation.

The snow had picked up but wasn't adding to the patches already on the ground. I rested, my elbows on my corral, and gazed out the shed's low mouth across the pasture that fell away to Lake Snowden. I looked southeast, up the frozen reach of water; the lake was a gray blur in the snow and dark at its edges with ice. The forested hills on the far shore were black, and smudged with the green of pines. The crooked barky posts of Massey's shed and the brim of its roof framed the view, the best on the farm, my new sheep barn. This is where Kathy had wanted to build. Instead, I'd destroyed the cabin to make room for a house. I told myself I was proud we hadn't torn down Massey's shed, but looking at the lake I felt certain that one day an expensive house would occupy this very spot. We'd merely delayed development. Although the farm was in Appalachia, it was still in America after all.

When I'd shown Mom this view, she'd said pleasantly, "I agree with Kathy." Yet I was proud of myself for my initiative: I'd *acted*; no more dreaming. Often, growing up, I had resented taking Dad's advice that I'd asked him for—and he only offered it if I did ask—but now I had no one to credit or blame except myself. Mom had seemed to respect the fact that I knew my mind. I wasn't being "needy," as she'd called me one night on the phone (about the same time that Bailey had cuffed me) for complaining— "whining," she meant—about Athens, the region, our upheaval. At first it had surprised me that she hadn't said more about my plan to remove the cabin. But although she was never a doormat, she was a woman of her generation: the man made big decisions, and the woman supported them even if they seemed crazy to the world.

I called the dogs and headed toward the farmyard, my lungs filled with fresh air and my head full of plans. I felt thankful for so much enthusiasm being released in early middle age. Impulsively I turned and stood looking back. Massey's rude shed, the handmade squat gray cube streaked by snow, looked beautiful. The dogs collided near my legs. Doty growled at Jack, a deep rumble, and she raced downhill with him flying after, snapping at her heels.

I felt ashamed of my overheated frustrations, which always seemed merited but which this timeless day revealed as self-centered and ungrateful. I heard a clear counterargument, an answering voice. *Yes, the terrain prevented agrarian or industrial success. But you have a beautiful farm. Put down roots, grow. You're lucky. Blessed. It's a sin to be unhappy.* Back in Indiana, Kathy and I were just two more moderately successful middle-class people. Here, a secret seemed embodied in this old farm and in this region—a secret we might be here to learn.

The future still felt fraught with risk, yet as I turned to follow the dogs again I heard the voice. *Remember this afternoon in Massey's shed.*

CHAPTER SIX

You meet some seemingly warm and fuzzy people who love animals
to a self-sacrificing degree in the livestock biz.

—e-mail to our former minister

I WANTED TO RENAME THE FARM. "LOST VALLEY" DESCRIBED THE BOTTOM
ground now submerged beneath Lake Snowden. And people in this region
of hills and hollows had called countless farms Lost Valley or Hidden Valley.
The farm itself *was* hidden, even secretive, despite the tree disaster and the
curiosity we'd aroused, but I wanted a name that meant something to *me*.

As a boy, I'd loved naming animals. At the age of five, living at Stage
Road Ranch, I called my blue parakeet Hattie; I have no idea where the
name originated, yet it still sounds perfect. A laying hen years later in Florida was Vinnie Polivar Simon Ficus, a grandiose moniker for a little red hen,
though Vinnie offset the effect. I'd developed the precept that one should
commune with an animal before bestowing its name.

In the case of our farm, I now wish I'd consulted Claire, who named with
creative gusto. My first attempt was prosaic, adult: Lakeview Farm. I rolled
it around in my head, ordered a custom mix of bluegrass seed called Lakeview Farm Blend, and saw it printed on the sack. Soon it sounded generic on
my ear, like a subdivision selling lots. Farm names are a genre that farmers

soberly respect, even as they sometimes choose whimsical ones. I decided our farm's name should either be thoroughly practical and descriptive—Gilbert Stock Farm—or evocative of the land.

One mild winter afternoon I sat on my truck's tailgate in the clearing at the heart of the farm. With the soaring tulip poplars at my back, I focused on the triangular hemlocks that flanked the lane into the north pasture. Water, trickling from the high field into the hemlocks' sequestered vale on the edge of the farmyard, kept the soil moist and fertile. Our house would face the evergreens' entwined, upward-swept branches. Eastern hemlocks, I knew, grew naturally in the region, their seed carried into Appalachia from Canada by the meltwater of glaciers; you saw them in remote hollows where streams rippled and rocks erupted from the hillsides. This domesticated pair must have been planted by Mabel and Kenneth Vaught. Their doing such a thing, and their choice of species, and the fact that the finicky hemlocks were thriving bespoke love, aesthetic appreciation, and the spirit of the place.

I decided to call the farm Hemlock Valley. On Monday when I told my boss, David Sanders—director of the press and a poet—he pointed out the death imagery associated with the word hemlock because of the poisonous herb. Looking up in wonder at our trees, I'd stubbed my toe on death.

Then I remembered a place the farm and its old trees reminded me of. In Georgia we'd lived near a spring that welled up in the flatland where pine woods had been given over to peanuts and cotton. The spring remained, surrounded by a fringe of oaks hung with Spanish moss. Locals all knew it as the best swimming hole in Lee County—possibly the best in the State of Georgia—deep and cold, not half a mile from the lukewarm Muckalee Creek that meandered along the eastern border of Stage Road Ranch.

One summer after we'd moved away, on our annual visit the teenage daughter of one of Mom's friends took me there. The rows of cotton surrounding the oasis were dark green, with heavy velveteen leaves that seemed to nod in the heat. Here, at age twelve, I sensed around me a lost landscape, a ghost world of longleaf pines and wire grass. We moved into the shade of the oaks and I stared at the water. The spring stared back, a blue eye. "It's bottomless," Susan said. Water bulged at the surface from a powerful submerged force, then spread and flattened. The silent upwelling was mesmerizing. And menacing, though its threat felt different from that of the alligators I imagined swimming nearby in the shallow muddy creek. The pool's unknown depth and the strange enticement of its pulsating force were what frightened me. My eyes got lost in the blurry swirl, and I couldn't see any footing beyond a narrow rim of sand. Looking into the aquamarine water, I resisted its pull and pictured myself pitching headfirst into the depths against my will.

This place, scary and sacred all at once, was known as Mossy Dell.

A decade after that visit to the spring, my employer, a newspaper editor in Columbus, Georgia, an hour west, jeered when I handed him my forwarding address. "*Coral Tree Farm. Ooh. Ooh,*" he crooned acidly about the name of Dad's farmette. That stung, though I knew he was miffed because I was leaving his newspaper for a bigger one, my hometown rag. In truth I was leaving to be near Dad's new farm, but it turned out I'd get too busy covering murders and space launches to spend much time there.

Dad's romantic name arose from an airy tropical plant his old aviation buddy, my namesake, had given him. Now that he had retired as vice president of Pan American's aerospace division, Dad's retirement project was to turn five acres of sand, Bahia grass, and live oak trees into a farm with a cash crop of some kind. Still thinking like a cattleman, he first built a pole barn and a corral, and even bought a head-squeeze before deciding that he wasn't up to handling steers. He explored raising pygmy goats for the pet market—his increasingly esoteric reading included *Aids to Goatkeeping*—and I was amused by the image of him leading a knee-high pot-bellied herd.

He had a pond dug and stocked it with wood ducks, so ornate they looked painted, in remembrance of the lake on his boyhood farm and its extensive collection of wild waterfowl. My mother, in fidelity to her own roots, tended a flock of white and gray domestic geese and began to raise chickens and guinea fowl. Dad wouldn't permit Mom to slaughter their poultry for their table, so the chickens' and guineas' only peril came from Florida's numerous hawks and raccoons. Mom bought an incubator that could hatch 125 guinea eggs at once, and she kept it humming, staying well ahead of the predators.

While Dad researched farm enterprises, he seeded buckwheat, and then, dragging a petite disk behind his little Kubota tractor, he chopped the plants into the thin soil to add organic matter. He built a trellis and planted muscadines, the big southern slip-skin grapes, as a potential market crop, though he didn't pursue that angle, probably because he couldn't produce enough fruit to sell at wholesale, and a retail business would've required dealing with too many customers. Mom, at least, loved eating the grapes. In winter, he sowed annual rye grass, which turned the farm into an emerald swatch in the khaki landscape of West Melbourne, nine miles from the Atlantic Ocean. His mania for generalized soil improvement vanished when he didn't need fertile earth for the enterprise he'd settled on. He would raise trees in plastic pots on top of the ground. He devoured books, took horticulture classes at the local two-year college, and visited growers.

Finally he specialized in native oak trees and wax-myrtle shrubs, and thereby he stumbled into Florida's native plant movement. Business at Coral Tree Farm was good: eventually he netted as much as $15,000 a year from his vest-pocket farm.

When I visited with my Labrador puppy from Cocoa, where I lived near the newspaper, I walked along rows of young oaks and wax myrtles that caught the sun in their thick glossy leaves. He grew untold thousands of the shrubs by picking their gray peppercorn-like fruit and then roughly rolling them to scarify their tough shells; he sowed the seeds in the shade of a lath house. I think he loved his oaks best. When he'd harvested all the acorns from the trees that shaded his farm, he collected seeds from oaks in Orlando, at Disney World, where Meg worked. "I've heard optimism described as an old man planting an acorn," he said. But Dad took the best care of his plants and got amazing growth rates.

When he wasn't tending his nursery, he sipped iced tea in the house, looking out the window for his customers, nurserymen and builders, who headed for the "Ring Bell for Service!" sign—white with sprightly red script—but who seldom could ring before he was at their sides.

After my fellowship year up north, I brought Kathy there to meet Mom and Dad. Kathy's encounter with him still looms as historic in my mind. It was 1983 and I'd just followed her to Indiana. At the Orlando airport, Dad spotted me with my lover at the top of the stairs exiting the airplane. He noticed Kathy looking for him, her brow furrowed, scanning the crowd for a man she'd never met.

"She's Type A," he said approvingly to me as we got our luggage. He was invoking his own personality, self-diagnosed from reading *Type A Behavior and Your Heart*. The bestseller described the hard chargers who got a lot done, defying the indifference of the world, yet who suffered from their own impatience and anger.

At Coral Tree Farm, Dad got out his Cranbrook yearbooks to show her. He'd never done such a thing, showed anyone the elegant 1930s volumes— certainly never to any girlfriend of mine. There were many mentions and photographs of him in *The Brook*, reflecting his popularity and involvement. He was in every Christmas pageant and several plays, and he participated on the football, tennis, soccer, and rifle teams. He was captain of the fencing team and led the glee club. Senior year, he was chairman of the Red Cross and dance committees, and served as head prefect, a post modeled after English student governance that involved maintaining morale, discipline, and loyalty, taking attendance at Chapel, and meeting weekly with the headmaster.

I was touched that Dad was putting on a show for Kathy. I suppose her status as a professor made his sharing of his lost world relevant, but he also was showing a pretty young woman the smoldering dreamboat he'd been. I think he sensed more about Kathy than their short acquaintance would suggest. I imagine his assessment happened like mine had, in the way that we know some people at once, or presume we do—how we see in a countenance qualities we admire and need.

Maybe, like me, Dad saw Kathy's strength and the way it was allied with, and softened by, kindness. Yes, she was sweet, without Type A anger, yet she'd launched herself into the world almost as fiercely as he had. Growing up on her family's farm in northwestern Ohio, she'd been a willful child. Not yet toilet-trained, she'd run away, angry or not, but searching for something. Her mother, Mary, would look out a window and see her fourth child receding, the diapered girl disappearing across a soybean field. "Jim," she'd say to her oldest, "go get Kathy." He ran, scooped her up; she kicked, but giggled too.

No, Kathy didn't have my fear—of drifting rootless, alone and unknown—but surely knew what must have been Dad's terror: of being trapped, unable to take off for the horizon.

The sheep breed I was interested in had an odd name: Katahdin. It was vaguely familiar to me from Thoreau's account of his climb of Mount Katahdin, in Maine, but I'd never said the word aloud. The previous year when I'd called Zinnia Rhodes about her ram, a boy who'd answered the phone had impatiently corrected my fumble at pronunciation. "It's *Kuh-TA-Din*," he'd said in disgust.

I'd been reading about Katahdins in my farming magazines—they were the only sheep breed even advertised in the *Grassfarmer*, with its cattle focus—and on the Internet. The shepherds who chatted daily on Sheep-L were grouchy traditionalists and seemed dubious, yet Katahdins were surging in popularity.

Most breeds have arisen slowly, emerging in a region over decades or even centuries, but Katahdins resulted from the vision of one man. Michael Piel, who lived in northern Maine, was a lifelong shepherd, the heir to a brewing-company fortune, and an amateur geneticist. He kept 1,000 head of mixed-breed ewes—a huge number for New England—and in addition to marketing their offspring, he bought as many as 3,000 weaned lambs a year to feed. One day in 1956, the Katahdin society's literature went, Piel was reading *National Geographic* and saw a full-page photograph of a man walking behind some bony ruminants on a rocky hillside in the Virgin Islands; the caption said, "Goats Browse Contentedly Where Sugar Barons Once Held Sway." The animals did have slick hair coats like goats, but Piel's sharp eye picked out their long tails, not the upswept deer-like tails of goats. These were sheep, not goats: woolless sheep.

In the sheep world, I knew from Dad's old farming books, breeds were ranked hierarchically according to fleece quality, which was expressed in terms of their kinship to the Merino, a Spanish breed that's the gold standard in fine-wool production, and the foundation of many later breeds.

By the 1400s Merino wool built empires. The hardy yet slow-growing and diminutive sheep had so much oily lanolin in their fine, dense wool that it flavored their meat.

Bucking history, Piel was intrigued by the idea of creating a new breed grown for meat alone, one that didn't require shearing. Synthetic fibers were growing popular and wool prices were depressed. And because his own ewes grew coarse wool and his fleecy lambs went to slaughter unshorn, 20 percent of his feed dollar was wasted growing wool. He contacted the head of the agricultural experiment stations in the Virgin Islands and asked about obtaining some hair sheep, which were said to have arrived in the islands from Africa during the slave trade. In November 1957 Piel imported three lambs from St. Croix. He bred the ram, a white animal he named King Tut, to ewes of traditional breeds as well as to the hair-sheep ewes, one white and one tan. Later, he mixed in other breeds, including Cheviots, a hardy Scottish hill sheep, and Dorsets from England's richer lowland pastures.

Ten years later, Piel was ready to unveil his work. His announcement—"Will the Sheep of the Future Be SINGLE-Purpose Sheep?"—appeared in a 1966 copy of *The Shepherd*. The magazine's editor felt compelled to add a cautionary note: "By his own admission, Mr. Piel's views are unorthodox, and highly controversial." Piel pointed out that wild sheep shed. Humans have selected domestic sheep for at least six thousand years to retain their wool for convenient harvest; hair sheep, most of them in Africa, now make up only 10 percent of the world's domestic flock. Like wild sheep, Katahdins grow thick winter fleeces, a mix of hair and wool, and in spring begin to shed, revealing their smooth hair undercoats.

Nothing against wool, but like Piel I wasn't interested in growing it. Its value had continued to decline, at least wool from typical farm flocks, and was a problem to get rid of. Mike Guthrie hardly received enough from it to compensate him for shearing. And no heir to Athens County's elderly sheep shearer was in sight.

Another benefit of Katahdins was that they were said to resist internal parasites, thanks to genes from the tropical sheep, which had survived under relentless nematode attack for generations. Mike and other shepherds in southeastern Ohio, as in much of the nation with warm humid summers, were dosing lambs with deworming medicine every three to four weeks. With their close grazing, sheep foraging in infested pastures swallow many parasites, and sheep evolved with limited ability to resist them. Sheep are native to the Middle East, in the vicinity of today's Iran, in mountainous terrain and high-elevation arid plains, ecosystems that limit parasite survival. Yet farmers have moved sheep—and their parasites—into every climate and terrain, even swamps.

Katahdins' other advertised virtue was prolificacy, yet another gift from the African sheep. Although experts had been telling American shepherds for decades that they needed to market a 200 percent lamb crop, an average of twins from each ewe—the first lamb to cover expenses and the second for profit—traditional breeds were lambing closer to 100 percent. Mike had already convinced me that profit lay in prolificacy. He even welcomed triplets, which many shepherds feared because of the labor of rearing some as bottle babies and because they're smaller. "They never look as nice," Mike allowed, "but they give me more weight to sell off this farm. I'm paid by the pound, not by how pretty they are."

Piel reported in the *Shepherd* that it was difficult to meld his parent breeds into a shedding, meaty animal, his three original hair lambs having been "as delicate and fine-boned as fawns." Less than ten years later, he gathered 120 of his meatiest, slickest ewes and called them Katahdin, after Maine's tallest mountain. He died too soon, of a heart attack in 1976, to see his breed expand when Heifer Project International, the livestock charity, adopted Katahdins in the 1980s and kept a large flock at its Arkansas ranch to produce stock for the Third World and the American South.

Katahdins looked like a way for me to raise prolific sheep that didn't need to be crossbred to thrive—they retained hybrid vigor from their diverse gene pool—and without having to deal with worthless wool. Maybe they'd need less deworming. To me, these sheep epitomized the American melting pot. And the heretical Katahdin seemed perfect for Appalachia.

In late December of that second winter I drove to Dayton under a pale sky to see my first Katahdins. The Goss family had raised sheep for generations on their farm and had shifted to hair sheep. They had a Katahdin ram listed for sale in Ohio's *Farm and Dairy* newspaper.

I found the farm's manager, Glen Fletcher, inside the double doorway of a big white barn, pitchforking manure into a spreader. He was cleaning out the cavernous building, one fork thrust at a time, in preparation for lambing. Glen wore his gray hair in a bristle cut and had a neatly clipped beard; he was my height, but muscles stretched the sleeves of his flannel shirt.

"We got a new ram lamb the other day," he said, leading me deeper into the barn. At the end of the aisle, a small chocolate-brown animal stood and stretched in a stall. He was curious and nibbled at my canvas jacket's knit cuff as we stood looking at him in the barn's dusky light.

"He's sure cute," I said.

"He's sure wooly," Glen said, leaning on the glossy handle of his pitchfork. "We've worked hard on the flock's coats. These are supposed to be *hair* sheep. But we'll see. Sometimes they're wooly as lambs and shed out as yearlings."

Glen ushered me down the corridor toward a bright rectangle of light, a Dutch door. Confined in a double stall behind white boards, the advertised Katahdin turned to face us. The low-slung animal had a slick white coat freckled with tan spots, red hocks, and black hooves; a thick, showy mane of coarse white hair draped his shoulders like a cape and crested at his chest. I thought he was beautiful and wanted him immediately. Sharing his pen was a distraction: an enormous wooly white ram.

"He's huge!" I said. "You could ride him like a horse."

"I have," Glen laughed. "He's a Montadale. We were using him for crossbreeding. But we're selling him and going with straight Katahdins."

"I like the look of that Katahdin. What does he weigh?"

"He'll go about two hundred," Glen said, appraising the ram. "How many ewes do you have?"

"Not any, yet."

Glen turned to look at me. "You're shopping for a *ram* before you have *ewes?*" He was grinning, a twinkle in his blue eyes, teasing like he'd known me for years.

"Well, I want Katahdins but hadn't even seen one. This is the first I've seen. Second, after your ram lamb. We're planning to get ewes next year. Are you going to be selling any?"

"No, just him." He explained that the farm's owners, Henry and Mary Goss, had started with six Katahdins and now had fifty. "Henry wants one hundred," Glen said. "He's sick, and I'm trying to get there for him. Would you like to see the ewes?"

Glen drove me across a pasture in his red Chevy pickup. Ahead of us, colorful sheep moved across the grass: ice-milk whites with black flecks, mahogany reds, many warm tans, and even a few chocolate-colored sheep.

"It's like a herd of Indian ponies!"

"I like the different colors," Glen said, gazing out. "It's fun after the white sheep."

"I'm interested in that ram, Glen."

"Good. We've used him for five years and don't have much left we can breed him to. I'd hate to send him down the road."

I nodded, understanding his euphemism for slaughter. "Let me think about it and call you," I said as he turned the truck back toward the barn.

Headed back to Athens, I hatched a plan. If the Gosses would agree to board the ram, this could work. And the ram was such a deal, only $100. As soon as I got home, I telephoned Glen and told him I wanted to buy. Then I mailed a check to Mary Goss. Then I worried about Kathy's reaction.

"First a dog, now a ram," Kathy said, standing at the stove stir-frying vegetables and meat. "It might have been nice to discuss getting another animal for once."

"We talked about Jack," I said.

"No, we didn't," she said.

"I thought we did."

"I see."

As I poured her a glass of wine, I could tell she was more interested than irritated—and this was part of the larger plan, admittedly mine, that she'd agreed to.

That Saturday, three days after Christmas, we drove with the kids to see our ram, whom I was calling Mister George, after a courtly family friend in Georgia.

At the Goss farm, Glen introduced us to Henry Goss's brother, Roy, a tall man in a blue nylon windbreaker. I got out my camera as we stood looking at the Katahdin ram, now alone in the pen, but it was too dark inside for pictures. Roy entered the stall, grabbed Mister George under his chin, and guided him out the gate and through the Dutch door. In the corral I snapped pictures of Roy grinning with the Katahdin, which he held easily, one hand on the animal's chest and one gripping his flank. I realized I didn't have a clue how to handle sheep. Kathy, Claire, and Tom stood in a wary cluster eyeing this animal. He was handsome, even dignified with his mane, but didn't look remotely cuddly.

After Roy and Glen returned him to his stall, we walked to the Gosses' farmhouse, a green two-story clapboard with black shutters and white trim. In the warm kitchen, Mary Goss, a heavy woman with white hair and silver-framed glasses, moved toward us using a chrome walker, smiling broadly. She told us she was eighty-five and Henry was eighty-seven. Mary stood talking with Kathy as I settled around a maple table with Roy, Glen, Claire, and Tom.

Roy said he'd grown up on the farm with Henry, but had become a journalist and, now retired, was visiting from Arizona. I returned to my agenda: getting ewes for Mister George.

"When you get your herd the size you want, will you sell ewes?" I asked Glen.

"Well, the future's unclear because Henry's so sick."

"Prostate cancer," Roy said. "He's back from the hospital. For good."

"But Mary wants to keep the herd going," Glen said.

"It's a nice-looking herd," I said.

"You two keep talking about the *herd*," said Roy with an exasperated smile. "A group of sheep is a *flock*."

I felt myself blush. Despite all my study, I'd muffed basic nomenclature.

Kathy took mugs of hot chocolate from Mary for Claire and Tom, and Mary clumped toward us from the stove. "Would you like to meet Henry?" she asked me. "He's just down the hall."

"Yes," I said, though I didn't want to meet him, to face a dying man with my blessings.

Henry lay on a wheeled hospital bed. Tucked under a white sheet and a quilt with red and blue patches, he raised his arm slowly to shake my hand, his narrow face gray. A pair of windows across from the bed overlooked the closely bitten pasture in front of the barn and admitted bluish light. The short winter afternoon was fading. I sat on a wooden chair, a sewing machine at my elbow.

"I was born and grew up here," he said. "Mary and I lived out west for several years. I was an electrical engineer. We raised our kids, kept the farm going long-distance when we had to. There's been sheep here for 150 years."

"I appreciate the ram," I said. "He's a beauty, and so is your flock."

"I wish I could be outside," he said, "working with the sheep."

On a lunch break that week, browsing in Athens's lone used bookstore, I discovered a new hero when I pulled from the dusty shelves a red cloth-bound book with *RFD* in gold letters on its spine.

Named after the 1891 designation that brought mail delivery to rural dwellers, *RFD* was a farm memoir set in the region just west of Athens in the state's frontier capital of Chillicothe. The author was Charles Allen Smart. During the mid-1930s, he was thirty and living an urban, intellectual life steeped in the arts when land ties pulled him to southern Ohio. Smart had published two novels and was teaching at a prep school in Connecticut when he inherited the family's ancestral home place upon his aunt's death. Although he'd grown up in Cleveland and later Long Island, he'd spent summers on the farm, named Oak Hill.

Here was another writer-farmer like Louis Bromfield. Given my own relocation trauma, it was comforting to read of Smart's culture shock, in a region he'd known his whole life. In those days "living in the country" was a meaningful distinction in America—especially in southern Ohio in the Great Depression. Roads were terrible, telephone lines spotty, and rural electrification was still news. If I thought Appalachian Ohio was remote in the late 1990s, the region Smart inhabited was truly blighted, afflicted by poverty, racism, ignorance, isolation. Neither northern nor southern in geography and sensibility, the region moldered in some lost netherland between.

And Smart was a Depression-era Red who dreamed of an "American Marxism." In light of his open radicalism, the fact that *RFD* had been a Book-of-the-Month Club selection amazed me. It showed how deeply America had been shaken by hardship. Smart wrote of his "baffled admiration" for his stalwart neighbors, pillars of the grange, and every one of them still a Republican. They clearly liked him, and he saw a touching solidarity in their rural fellowship, though he called himself and his Eastern wife "freaks" who alarmed the provincial folk and upset the American Legion by staging Clifford Odets's play *Waiting for Lefty* in Chillicothe Little Theater.

The farmers signed him up as an Ohio Farm Bureau member anyway, while nodding politely over his handwringing about the bureau's ultraconservatism. (I planned to join too, probably for the same reason Smart had: the Farm Bureau's validation and its discounts on products and car insurance; it was really an insurance company.) To his chagrin, at his first meeting Smart breached rural protocol by sitting down at the women's card table.

Bromfield, returning to Ohio from France the year *RFD* was published, surely noticed that a kid just to the south had managed to wring a bestseller from a rambling farm memoir. He'd also have noticed Smart's glancing dismissal of "Mansfield stories," an allusion to Bromfield's operatic condemnation of his provincial hometown, Mansfield, Ohio, which Bromfield saw as an agrarian utopia defiled by sooty factories and grimy tenements ringing with the voices of foreign workers. (Smart preferred stories by Chekhov and Turgenev.) As if in response to this leftist aesthete, Bromfield cranked up his formidable narrative skill and, five years later, produced *Pleasant Valley*, his first nonfiction book. Thus Bromfield began his rebirth as an important chronicler of ecological farming.

There wasn't a copy of *RFD* in my father's extensive farm library, and I wondered if he'd read it, maybe in the wartime edition issued for servicemen overseas. Dad had a conservative's sober view of human nature and would've found Smart's socialism repugnant. Nor would he have been charmed by Smart's subsistence activities like hand-milking a cow while singing Gregorian chants. Within the romance of Dad's own choices, he tried to be businesslike, which also happened to be the new postwar farm model: specialization. Smart, wary of farming's growing relationship with chemistry and industry, even took a swipe at Dad's brave new world of hydroponics, looking askance at vegetables from "tanks of chemicals." Disgusted by all forms of "buying and selling," Smart saw farming as a way of life above all.

I learned that *RFD* had gotten pretty good reviews in 1938. (In a snippy notice, *Time* magazine noted that Smart's desired income was $3,000, while the average family in his Ross County was getting by on only $572 a year.) I decided to lobby for Ohio University Press to republish the forgotten story.

The director and the senior editor, Gillian Berchowitz, and I had been talking around the water cooler in the fall about our feeling that Ohio University was refusing to acknowledge its region. *RFD* might be our reply.

I was bleeding on that January afternoon as I sawed away on a block of wood clasped in my cheap vise on a wobbly workbench beside the furnace, downstairs off the family room. I'd scraped my knuckles and cut my right forefinger. I was trying to transform the chunk of dry white pine into a car for Tom. He'd lost interest in the slow process and was watching TV with Claire.

I'd enrolled Tom in Cub Scouts, and the Pinewood Derby model car race was the Scouts' sexiest annual event. Held in February, it excited Tom and was driving me to renewed efforts to earn Good Dad status. The derby was a race of colorful cars, which gravity hurtled down a sloped track. Tom liked the pretty racecars, but he really wanted a shiny golden trophy.

"Look at your car," I said, plopping onto the couch between Claire and Tom. The piece of wood was finally wedge-shaped, though lopsided. It sat in my lap on a sheet of newspaper; I would try coarse sandpaper on it next.

"Nice job on *Tom's* car, Dad," Claire said.

"Fathers help," I said. "We have to do the cutting. Look at my hands. Tom's going to paint it and help me put the wheels on."

"You're bleeding on the couch."

It was worth some blood to redeem myself as a father; I'd struggled to help him earn merit badges, and on our first campout, on a rainy weekend the previous November, I'd felt inept. In Athens, scouting was a male province, but it favored men who were avid outdoorsmen already, or who had a geeky interest in its clothing and gear.

In the suburbs of my boyhood, mothers ran the Cub packs because their husbands were away at the Kennedy Space Center even on weekends. My troop ate a lot of cookies and earned few merit badges. And without Dad's or any man's help, racing was beyond my wildest fantasy. Even seeing Pinewood was impossible. So, scouting, and Pinewood Derby, with Tom.

And karate. After the kids' early experience, I'd found a studio less grimly combative, but Claire quit. To keep Tom going, I'd suited up and we took lessons together a couple of times a week. He soon outranked me, a green belt to my lowly yellow, and was steadily pulling away. I hoped the lessons would make him more physically confident, less a target for bullies. I didn't want to know differently—what else could I do? Bullies were part of growing up. I'd been bullied myself until I'd learned to fight back. "Your father's going to teach you to protect yourself," Mom had promised me once after two boys had singled me out for daily abuse. I waited for his help, but Dad was busy.

Now I was. I felt I'd been a good father in Indiana and feared I wasn't anymore; my unstructured downtime had disappeared, plowed into Mossy Dell. I saw myself helplessly repeating my distracted father's pattern. Like most men of my generation, I could still say I was a better father than my own. But Dad had set the bar so low.

One evening I'd listened as Tom recapped the plot of a novel he was reading about a man and his boy who wandered the English countryside together in a quaint homemade trailer. Their lives were intertwined. The plot was fiction and surely a fantasy of what the writer had desired from his own father. But it pained me to hear Tom's simple summation: "His father had time for him."

Scouting for ewes, I drove Kathy and the kids across West Virginia one Saturday in our van. The mountains, so lushly green in summer, were brooding gray battlements in January. In the lonely western tail of Maryland the highway pushed through bleak mountain passes. We continued east, descending, and began to see barns, houses, and bare-mud crop fields. Near Frederick

I took a winding secondary road into the countryside, then a gravel road through a valley that brought us to our destination: Simon Sez Farm.

Jo Simon, a petite blond woman, opened the door into a room above a concrete-block two-car garage. Ted Simon, pear-shaped and ruddy as a skier, grinned and shook my hand and returned to his Laz-E-Boy. Cardboard boxes, grocery sacks, dishes, and appliances crowded the sitting-room floor, covered an adjoining dining table, and spread onto the kitchen counters. Jo darted to the table and swept clutter from chairs so that Kathy and the kids could sit; I settled on a plaid couch beside Ted's leather recliner. A coffee table in front of me was heaped with unopened mail.

"Excuse our mess," Jo said. "We moved last year, built this place, and then rented the old farmhouse. Next year we'll build a new house and rent this."

Kathy looked meaningfully at me as Jo spoke. "We moved a year and half ago," Kathy said, "and still haven't unpacked." She added diplomatically, "You've really been through it."

Jo gestured to a framed image hanging on the wall over the table. "That was our farm," she said. The aerial photograph showed an attractive brick house overlooking a pond; board fences painted black enclosed the pastures. The manicured acreage was eerily reminiscent of our Indiana place. "Tim wanted more pasture for his beef herd, and I wanted more land for my sheep," she said. "Everything close to D.C. is priced for development. So was this farm, but the heirs were fighting among themselves. The house was practically unlivable, so we built this."

"We'll get things under control," Tim said. "Jo commutes two hours into D.C. every morning, and I cover parts of three states in sales, so we're gone a lot. The flock was up to seventy ewes before the move, and she's cut back to thirty. We've got the land now, not the time!"

Jo had said she would sell me Katahdin ewe lambs, purebred but unregistered, for only $100 each. The Simons focused on practical meat production, on working animals, according to a man who sold the Gosses rams and who got $225 for registered ewe lambs. When I'd called him and asked for a lead on commercial stock, he'd mentioned Jo and Tim. He indicated that they were serious breeders yet unconcerned with bureaucratic registration papers. This fit with a maxim I'd developed years before: Beware of rich men selling dogs. I'd once given my parents a Labrador puppy that a veterinarian had sold to me for three times what he'd paid its actual breeders, which I learned when I tracked them down for training advice. I had since encountered other people who hawked registered animals for what seemed unwarranted prices. As Mike Guthrie liked to growl, in a neat summation of practical livestock-breeding theory, "The only reason for purebreds is to make crossbreds."

I produced a photograph of Mister George, eager for Jo's reaction to my ram. "He was bred out in Kansas, from Heifer Project stock, by Laura and Doug Fortmeyer," I said. She settled beside me on the couch and held the picture at arm's length. Her pale face and sharp features exuded tense vulnerability; beyond, Tim smiled in his recliner, feet up. At the table, Kathy said something to Claire, and Claire shook her head and crossed her arms.

Jo didn't speak, just stared at my photograph.

Finally I reached for the snapshot and Jo distractedly handed it over. "The Gosses really like him," I said as she rose and headed for the kitchen. Her silence mystified and embarrassed me. Did George look that bad? Was his conformation poor? I had little inkling then about breeders' rivalries and plain snobbery, different goals and divergent farming practices. A hobby shepherd, whose sheep were named pets and who belonged to the breed association for fellowship, felt slighted by a major breeder who had nothing to say to him. A working farmer who made her living from the land resented part-timers who sold stock below cost.

This world was just like any other; in fact the breed society reminded me of a church. We worshipped the mighty Katahdin, yet the manner in which our devotion was expressed made the difference. When we'd talked on the phone, Jo had seemed proud that her ewes descended from Maine Katahdins; she was planning to travel for her next ram to the Piel Farm, the headwaters of the breed, still operating after Piel's death. In purchasing my ram I'd thrown in with a different wing of the congregation, the westerly Heifer Project flock. From what I'd heard, the Heifer Katahdins were larger and more vigorous than the pure Maine bloodline. Perhaps to Jo, Piel sheep remained the only true Katahdins. Then again, maybe she was just tired and frazzled.

In the kitchen, she prepared bottles for lambs, and we left Tim in his recliner, still faintly smiling, and trooped to the frigid barn. Upon hearing Jo's voice, lambs ran from the recesses of the dark building. They had thick, wavy white coats and looked like any other lambs to me—but then, I'd never seen young lambs, Katahdin or otherwise. Claire and Tom held bottles, their eyes widening in surprise at the lambs' fierce nursing; the small creatures pummeled the rubber nipples—thrusting with their necks, digging in with their small hooves, and bucking forward—and backed the children against a wall. Jo's ewes had lambed even earlier than the Gosses', her flock's birthing having begun on Christmas Eve. I didn't see many of her ewes, which avoided us, and didn't remember to ask to see *her* ram. A storm front was forecast, and we said our goodbyes, hoping to avoid hitting heavy snow in the West Virginia mountains. "I'll send a deposit on fifteen ewe lambs," I said.

Driving home, Kathy and the kids silent, I felt very alone with my sheep plans.

"Their place was kind of a mess," I ventured.

"I'm *amazed* at what people put themselves through," Kathy said. "To farm. For animals."

I absorbed this silently as veiled criticism of me. I was behaving more like Jo and Tim than like practical Kathy. I knew she wanted me to wait to acquire sheep until we were living at Mossy Dell. That made sense, and yet I wouldn't consider it. I'd slow down later.

I gave the van more gas as it climbed; the sky was gray over the rising slopes. Since high school, I'd wanted to do something challenging, maybe even dangerous. Instead, I'd worked, all my spare time and in summers. From ages sixteen to eighteen, I bagged groceries and stocked shelves at Winn Dixie. In college, I attended classes even in summer, and had worked as a clothes salesman and a janitor. Graduated to the world of adult work, I'd jumped from one newspaper job to another. Kathy and I took our first vacation after we were parents to Claire and Tom. We'd gone to Pike Lake State Park, in Appalachian Ohio not an hour from Athens, the only place with a cabin on short notice; we'd hustled the kids off the beach when we saw a woman washing the stump of her leg in the water.

Now, fetched up in an Appalachian backwater without goods or services, but where folks seemed to follow their hearts with hardly a doubt, I knew what I wanted. *No more idle dreaming.* Like Dad, I'd act.

CHAPTER SEVEN

Of course, the proof is in the pudding.
    —Mom's reply to my assertion that I
        could make money raising sheep

BY FEBRUARY WE'D SPENT COUNTLESS HOURS AND SEVERAL THOUsand dollars planning a house, tearing down the cabin at Mossy Dell, and preparing the site. First we'd developed a floor plan with a national company that sold custom kit homes, airy timber-framed structures, before backing out over the cost. Then we'd worked with a local builder, who tried to approximate the timber frame's floor plan on his computer, but created a stark rectangular box. Finally we'd hired a local architect to draw a house fitted to the sloping site. At last we clutched blueprints and were ready to get bids from contractors.

As a diversion, I e-mailed my skeptical friend Bailey for advice on buying a canoe to throw on Lake Snowden in the spring. Claire and Tom could play with the boat like a big toy in the farmyard pond. Late winter also was a time for researching and ordering plants, for picturing them growing in various spots. We'd moved into a region with milder weather, which broadened planting possibilities and surely accounted for many of the differences in flora we'd noticed. Bloomington was in the southern part of zone 5 on the official plant hardiness map of the United States Department of Agriculture, while Athens was in the northern half of zone 6—a slight difference on the map, but the winter felt much less severe. We could grow everything we'd grown in Indiana, and we could try new species, like southern magnolia. Planting *Magnolia grandiflora* was risky, but I'd spied the glossy broad-leaved evergreens along the Ohio River, which moderated the cold. I ordered a hardy variety, a wild magnolia strain that the catalog claimed had been found growing in southern Indiana.

Mossy Dell's farmyard, already diverse, a blend of wild and domestic, presented boundless planting opportunities. Wholesale creation was unnecessary, but I could introduce new species and fill gaps created by the windstorm.

A receipt from J.W. Jung Seed Co. shows that I ordered four climbing roses of the heirloom variety Zéphirine Drouhin, which was named in 1868 for the wife of a French rose enthusiast. I hadn't grown roses, but I thought Kathy would like them, and romantic rambling roses would embellish the farmyard. Zéphirine Drouhin was said to tolerate shade and to produce fragrant pink blooms; I'd plant her along the stable's old corral fence. Claire adored raspberries, so I ordered eighteen plants. And twenty asparagus starts of two varieties. I had liked the narrow, swordlike foliage of a Siberian iris I'd grown in Indiana, so I ordered the old standard, Caesar's Brother, which bears intense blue flowers; he'd be lovely beside our pond.

I tried to balance my selections among ornamental plants, food-bearing plants, and those for wildlife. Chestnut oaks, which grow on hillsides and bear sweet acorns for wildlife, seemed perfect, and I ordered several. A farmstead needs lots of lilacs, and I researched them, finally choosing Miss Ellen Willmot to join our farmyard; her white blossoms would light the clearing and perfume the air. Receipts were piling up at a worrisome rate, but I couldn't stop. I ordered black gum trees, chokeberry, shrub dogwood, bald cypress, winterberry holly, American holly, persimmon, and hemlock. I already had a nursery behind our house in town, but I ordered so many plants that in spring I'd need to stake out another nursery above Mossy Dell's stable.

Not to forget sheep, I researched the portable electric fencing that was helping America to join the revolution in grass farming. By then I saw that I'd never make Lost Valley's tired, overgrown fences sheep-worthy—not by the time Jo's ewe lambs were ready. I placed an order for rolls of something called Electronet, a mesh of plastic struts and strands of wire, supported by lightweight plastic fence posts. With electricity coursing through the flimsy barrier, the sheep would respect their boundaries, and coyotes would run yelping back into the hills. The fence could be rolled up, thrown over my shoulder, and carried to a fresh grazing area; it could meander around an orchard, trace the curve of a pond, climb a hill, and descend a gully. Fed a current of electricity through a glorified extension cord, an Electronet enclosure could float in the middle of a field like a lily pad on a pond. Without electrical power after my destruction of the cabin, I bought a $400 battery-powered German fence charger and two deep-cycle batteries; I could recharge one in town while I lugged the other out to Massey's shed.

As plants and equipment began to arrive, my canoe dream faded. I couldn't justify the expense of ordering the Old Town boat Bailey had recommended. There was so much to do for the flock, and I was running out of time.

On a cold Saturday afternoon in February, Tom and I joined the Cub Scout packs in East Elementary's gymnasium for Pinewood Derby. Winners would advance next weekend to regional competition at the mall.

Tom placed our car on the black racetrack and adjusted it so it pointed straight down the slope. He'd painted the car in Ohio University's colors: apple green body, white racing stripes. He placed its nose against a peg that rose through a slot in the track and lifted his hand away with elaborate care, like setting a mousetrap. Tom gave a little jump as he stared at the waiting car. I stood at the bottom of the track with other fathers and sons who milled around. Tom's opponent, a skinny kid with black hair, slapped down his car crookedly, and its plastic wheels snapped against the track. My hopes jumped. Obviously these races were up to gravity, so small tweaks surely made the difference.

When the peg dropped, the kid's purple racecar flew down the track, wheels clacking. Tom's rolled slowly behind—in comparison it barely moved. I stared, uncomprehending, as Tom retrieved our losing car and the announcer called the other kid's name.

"What did we do wrong?" I asked an older man dressed in a Scout uniform, his tan shirt embellished with red badges.

"Did you weight it?" He spoke from the side of his mouth, looking at the track.

"Weight it?"

"Yeah. Cars are allowed to weigh so much. After you carve them they're a couple ounces below the limit. So you add weight, little pieces of lead."

How embarrassing. I hadn't seen anything in the instructions about adding weight. I was barely on nodding terms with our den master and didn't really know the den's other fathers, who might have told me. The competition was between men, with boys as their surrogates—that's the way this game was played—and I stung with shame for letting Tom down.

Tom held his car against his chest and looked at two cars streaking down the track. "We'll do better next year, Bud," I said.

One March night after dinner, the phone rang. I asked Claire to get it, but she was halfway down the stairs to the TV room, and Kathy was running an errand with Tom. Stuck, I picked up. "Hello, neighbor," said Fred Paine's mellifluous voice. I could almost see him smiling in his smarmy way, leaning slightly forward for the kill: "Dolores and I were just talking here. We thought you might want to buy our place."

"The cornfield?"

"Our house here on the hill, too. We're having a new house built over on Radford Road. You probably didn't know that. Dolores wants a new home, for our retirement. Well, I'm not going to stop working, but I am slowing down."

He'd taken me by surprise. I struggled to form words, to keep my voice neutral. "I'm interested in the cornfield."

"It's all or nothing, neighbor. A package deal: $250,000. The field and the house."

"How much land are we talking about, Fred?"

"On your side of the road there's about thirty-five acres. Twenty are

tillable, the cornfield. And there's about fifteen in woods along your southern border. The house is on eighteen acres. I'm keeping the field to the
south of me. My boy helps me farm it with our other land. We drive all
over the county nowadays, planting here and there, ha ha. You know how it
is, neighbor."

"Yeah, I know how it is, Fred."

I hadn't been inside the modest brick ranch on Ridge Road, though I'd seen
it often enough; it marked my turn onto Snowden Road, up there atop the
tallest hill for miles. What I remembered from last June, when I was there
buying Fred's post pounder, was his littered farmyard and swaybacked tin
sheds. Still, this was a chance to gain a vast pasture and to protect Mossy
Dell from a subdivision at its entrance.

Kathy said she'd look, and so, six days after Fred's call, I drove our van
between the two black lions and crunched up the gravel driveway, admir-
ing the homestead's dramatic placement, so different from Mossy Dell in
its secluded valley. We climbed between facing rows of apple trees; above,
on the hill's peak, the house's red brick walls and low white roof showed
through the leafless branches of two large maples.

I parked and took my first close look at the house as we walked toward
it. Clearly it had been born from a stock blueprint and intended for a suburban street. The formal front door, meant to open onto a sidewalk near a
mailbox, faced Ridge Road across acres of fenced pasture. The entrance that
people actually used was off the driveway, through sliding-glass doors. A
papery gray hornet's nest the size of a sheep's head hung over a second set of
sliders. Off the rear of the house was a small addition, a room sided in white
vinyl. The lane past the house was lined with sheds and ended at a large tin-
sided pole barn. The gravel glinted with neon-green cans of Mountain Dew
soda, flattened by machinery, and red plastic shotgun shells. To the north
the hilltop overlooked a neighboring field, brushy and abandoned, dotted
with junked vehicles.

Fred slid open the door. His long face beaming, he emitted a booming
laugh, as if we were sharing a delicious joke. The house was dark—*such
tiny windows*—and smelled funny, sweetish. "This was supposed to be the

garage," he explained. "But when we were building, we thought it would make a nice family room. And it does! We put these glass doors in the garage-bay openings. The wood stove in here keeps it nice and warm."

Dolores came down a hall toward us, drying her hands on a blue dish towel. "Lord Almighty," she said, apropos of nothing.

"Built it myself in 1973," Fred said. "That's only twenty-three years. Come on, I'll show you around." He charged down the hall. We climbed a concrete step and walked to the kitchen, where Dolores busied herself at the sink. Kathy and I squeezed into a small dining area, and then Fred opened a door opposite the kitchen and we entered a formal living room, as airless as a funeral parlor, its plush carpet the color of lime sherbet. In front of us a picture window framed one of the bare maples; the nominal front door, to the right of the window, opened onto a concrete stoop above the pasture. Next, Fred showed us three small bedrooms at the south end of the house; as elsewhere, he'd covered the walls in glossy, plastic-looking wood-grain paneling, further darkening the interior.

We followed Fred's sagging khaki pants through a storm door off the dining nook and into the brick house's boxy white-vinyl addition. "This is my trophy room," he said. The head of a six-point buck stared down at us from the paneling; a coyote glanced at us in mid-stride from the floor below; on the opposite wall, a green bass arched its back and turned a gaping mouth toward us. I realized that the odd machine atop a paneled counter, which stood in the middle of the room like a bar for serving liquor, was a bullet press.

I longed to be outside. A small window overlooked the back yard, dominated by a submarine of a white propane tank. We saw the rear of Fred's tin shanties; the yard ended in two giant galvanized grain silos squatting on concrete pads. There wasn't a shade tree in sight. Evidently the view of the house from the road was the idea: the picture of a prosperous country estate, lions and all.

Fred led us back through the family room toward the basement door, and Dolores, following a few paces behind with her dishrag, mumbled, "I've gotten behind on pumping out water." As I descended behind Fred, I was surprised to see that about half the basement was flooded, even though no rain had fallen for two weeks. Kathy and I exchanged a look. Apparently

Fred didn't run a drainage pipe around the foundation to collect moisture because the top of the hill was high and dry. Presumably in this way he learned that a hole, wherever one digs it, fills with water. He'd drilled weep holes here and there in the cement-block walls to relieve water pressure. But the concrete floor, which rolled unevenly under our feet, sloped *away* from the drain. This explained the portable electric water pump and the tangle of extension cords and hoses.

"We've gotten a little water this winter," Fred said, chuckling absently. "Downspout extensions should take care of it. I've been meaning to put them on." I gazed at a chest freezer beached near the stair landing; a rusty steel office desk heaped with junk was shoved into a dark corner.

Fred and I walked down the driveway while Kathy wandered again through the house. "We're going to have an auction and get rid of some of this stuff," Fred said as we passed the sheds. The one nearest the house was a garage, with a tall overhead door for his dump truck; the next, open-fronted, where he'd kept the post pounder, was littered with bark and chunks of fire-wood. Past the grain bins, the shed nearest to the barn had partly collapsed, its roof having blown off. At the barn, he flipped a light switch—proof of a real feature: electricity—and illuminated a shiny red electric fence charger affixed in one corner.

"Does that stay?" I asked. Having bought a charger already for Mossy Dell, it would be nice to get one thrown in.

"It *stays*," he said with one of his oversized grins.

"That was depressing," Kathy said as we drove away. "All that dark paneling."

"He sure saved money on drywall . . . He doesn't know how strange we are to be interested."

"I'm *not* interested."

"But, honey, we could sell it, keep the cornfield, and still build at Mossy Dell this spring."

To her doubtful pause I added, "Look how much we'd get for the price of a house. It might take us longer to get settled over there, but we'd gain so much land and so much protection. Remember, he sold off those plots along the other side of Snowden Road."

"What about zoning?"

"What zoning? Mike Guthrie said there aren't even building codes around here."

We had to decide what to do quickly or we'd miss the spring selling season on our house. This felt like another emergency, an opportunity to seize or a loss to suffer. Of course, Fred could keep farming the field and unload just his house. But, if so, there was no way of knowing whether I'd ever have a shot at buying the land again.

That night I telephoned my practical, clear-eyed mother for advice. If I was determined to try for more land—an option she didn't endorse—her advice was to make a modest offer. "Lowball him," she said. "He doesn't have any money in that place. Anything he gets is gravy. You can get it cheap."

Two days later I called Fred and offered him $175,000, well below his asking price but above the $125,000 my mother advised. He just snorted.

At work the rest of that week, surrounded by manuscripts to publicize, I found it hard to focus on anything but my desire for Fred's cornfield and my worries about our impending sheep flock. I realized I'd neglected to ask Jo about the health of her flock, though she'd volunteered that her sheep were free of foot rot, a crippling, malodorous contagion that causes sheep pain and shepherds heartbreak (1.4 million hits on Google). Foot rot is widespread and difficult to eradicate, because sheep can carry the disease walled off in their hooves in dry periods without symptoms.

I hoped Jo was telling me the truth.

I refrained from e-mailing Bailey about my worries because I knew he'd call me an idiot.

Instead I concentrated on my research, first focusing on the foe I knew I'd face: internal parasites. Even though Katahdins were supposed to be more resistant to the pests than wooled breeds, such genetic protection is variable. Scientists estimate that 20 percent of sheep in a flock are magnets, hosting as much as 80 percent of the flock's parasites. The worst nematode species I'd face, common in areas with warm, humid growing seasons, was the barberpole worm, *Haemonchus contortus*, so called because of a red and white spiral visible through its translucent body. The red color was from sheep blood, its food, the white from its incredible payload of eggs: ensconced in a sheep, one adult female could lay 5,000 to 10,000 microscopic eggs daily. In one Ohio study, I read, forty-six lambs dropped more than *100 million* worm eggs *each day* onto the pasture. The eggs, encased in the sheep's manure pellets, could hatch in less than a week in warm weather.

The more I learned about *Haemonchus*, the more it seemed like a predator out of science fiction: tough, committed, creepy. Its legions of sightless larvae climb grass blades in droplets of dew and wait to be eaten. A tough sheath protects them as they linger, helping them survive on pasture in the shade provided by long grass. Three quick weeks after being consumed, the larvae mature, mate, and the females begin laying. The adult threadlike parasite is large as sheep roundworms go—up to an inch long—and uses a hook on its head to scrape the sheep's stomach and cause bleeding. Its saliva contains an anticoagulant to keep the blood flowing. Every time one of the parasites detaches and wriggles to a new feeding spot, the first point of attack continues to bleed.

Horrible. But even so, it puzzled me how such small creatures could cause enough blood loss to kill sixty-pound lambs and sometimes adult ewes. Then I recalled the worm's stunning fecundity, which translates into unending waves of infection. A ewe grazing summer pasture might consume as many as 30,000 larvae daily. If she hosted only 5,000 worms, she might lose a pint of blood weekly.

Lambs needed to be drenched relentlessly and on schedule. Despite the expense and labor involved, the prices being paid for lamb made such a system profitable. Shepherds on Sheep-L who reported having no worm problems, or who claimed minimal drenching, I noticed, were stocked lightly. Maybe they had cattle helping to clean up the pastures by serving as fatal hosts for specialized sheep parasites. A diversity of farm species might help explain how Ohio supported 8 to 9 million sheep during the Civil War (now there were fewer than 5 million sheep in the entire United States—Great Britain had as many). America's sheep also may have lost some genetic resistance to parasites, and shepherds probably had forgotten past knowledge about controlling parasites without laboratory-concocted dewormers.

But adding cattle to my sheep pasture wasn't a simple solution. For one thing, a disease that caused cows to waste away and die, Johnes, could be passed to sheep. Perversely, sheep had their own species-specific strain of Johnes as well and could catch either—or both—strains. And Johnes (pronounced Yo-nees) might be the cause of Crohn's intestinal disease in humans. At least some in the livestock world believed so. Widespread in U.S. dairy herds, Johnes could survive milk's pasteurization and sicken a human with a genetic proclivity to it. This theory hadn't reached the mainstream media. Neither had scientists' belief that the "mad cow" disease epidemic in Britain had been caused by rendering sheep ill with the wasting disease scrapie and feeding their powdered carcasses to cattle as a protein supplement. Such agribusiness practices had been outlawed, and shepherds were required to report suspected scrapie; infected flocks were being destroyed. But the U.S. government and agribusiness seemed to be doing little to eradicate Johnes, unlike in Europe where a concerted effort had effectively eliminated it.

Few exercises are more sobering than reading about such "zoonoses," diseases that can be transmitted from animals to humans. Zoonoses are common because of the long coevolution between people and domestic animals. Most major infectious diseases, including smallpox, measles, tuberculosis, and most famously AIDS, have jumped to humans from their origin in other species.

Newbies on Sheep-L barraged the jaded veterans with questions about how to avoid buying a disease with their animals, and how to eradicate one and achieve a "clean flock" if they did. Foot rot was the only disease Mike Guthrie had warned me about. He had little patience when I fretted to him about others. "You aren't going to have sheep without sheep diseases," he growled.

"I'm not going anymore," Claire said from the back of the van as Kathy drove us away from the Methodist church just off the Athens bypass.

"Me either," Tom chimed in, emboldened by his big sister.

Kathy and I didn't respond. I had spiritual yearnings, and we both valued the intentional community that a church represented. We missed our Methodist church back in Bloomington. Yet we'd just dragged the kids to

Athens's second Methodist church. And church-shopping, we saw too late, was a fatal error for keeping Claire and Tom involved without protest. We should've stayed at the big church downtown. But we hadn't felt welcomed there and had left after the minister and his assistant, dressed in trench coats and wearing sunglasses, had played the theme from *Mission Impossible* one Sunday on a tinny cassette recorder. Linking a TV show's theme to Christianity surely wasn't the worst idea ever broached from a pulpit, but we'd squirmed in our pew over their hokey extended dramatization.

"Did you know that was a Promise Keepers church?" I asked Kathy at home as we changed clothes.

"Of course not, Richard. Let's try the Presbyterian church downtown."

Doty, on alert since I'd started putting on farm clothes—fleece shirt, heavy canvas pants lined with flannel—jumped toward me as I tugged on boots. She stared at me so intently with her big brown eyes that I laughed. "Not today, girl," I said. Jack, always jealous of Doty's excitement, began biting her hocks. I grabbed him by the scruff of the neck, jerked him up, and shook him. "Stop it, you little pecker!" He dangled from my fist, his teeth bared in a grimace, looking like a furious possum. "I guess I'll take them," I said to Kathy, "run some meanness out of Jack."

"You need to go mope," she said.

"I guess. I'm tired of the Fred roller coaster." She nodded and narrowed her eyes in appraisal. *I* had returned us to the carnival with my negotiations.

As I drove, I wondered if we still were bargaining with Fred. If not, should we list our house with our realtor? Building at Mossy Dell would take a year out of our lives. And if we did build, would our magical farm be ruined by Fred's newest development at its entrance?

At the farm, I let the dogs out of the cab and watched Jack chase Doty into the woods at Mossy Dell's entrance. I slouched, hands in my pockets in the middle of the farmyard, trying to think of something to accomplish. It was mild for the last Sunday in February.

Fred Paine pulled into the farmyard in his black pickup. He must've seen me from his hilltop.

"Howdy, neighbor," he said, slamming his door. He tottered to the maple stump on the hill in front of the cabin site and sat down.

"Hey, Fred." I backed against the hill and slumped onto its face.

We both knew this was no mere visit, but the prelude to serious negotiation. I hoped my dejected aspect on the ground would strengthen my bargaining position. Context was everything. He'd weakened himself by rushing over. In turn, he knew my weakness, which was the ferocity of my desire for his cornfield.

"We'd like to sell our place to you," he began, "but your offer wasn't serious."

"We're serious. But we paid too much for this land. We're kind of strapped, Fred."

That was partly true—we'd paid a lot for Mossy Dell—but with two good jobs we could afford a full-price offer. I wasn't a businessman, let alone a decent bluffer, and wasn't sure how long I could keep this up. *Should I insult his house?* I couldn't.

"They've stopped making land, you know," he said. "When I moved here from West Virginia, almost forty years ago, I didn't have anything. Dolores and I put together our farm, one dab of dirt at a time. One here, another there. When a chance like this comes along, two great parcels, a fellow ought to take it."

I was silent for as long as I could stand it, which wasn't very long. "I'll talk it over with my wife," I said. "See what we can do."

At the end of the week I called Fred with a new offer—$225,000—and we waited. The bid was $25,000 below his asking price, but far more than we should have offered, given our reservations. Kathy had reluctantly agreed to my plan: fix up the house a little, move in, sell it, and build at Mossy Dell in another year or two.

Two days later Fred called. "Your offer's still low," he said—I felt something drop through my chest to my feet—then he finished, "but we'll take it."

I swallowed and paused before responding. Because of his house, I felt more sobered than elated at getting the cornfield. I'd bitched about our house in town, but Fred's was far worse—not just modest, but tawdry and neglected. And yet, this felt like a test of my tenuous belief in myself. I didn't want to live my whole life as a dreamer. I wanted to *do*.

"Good," I said.

Kathy called our realtor and asked him to get our house on the market quickly. It was late March, and only financing stood in the way of our owning two farms in Athens County. We'd close the sale in July so that Fred and Dolores could move into their new house. In the meantime, he agreed to dismantle the grain bins in the backyard and remove junked machinery and the larger piles of debris.

Fred also said that I could go ahead and plant the cornfield—now just mud, stubbled with the gray, weathered butts of cornstalks—instead of leaving it bare until midsummer. I'd fantasized about improved grass species and the clean slate the field represented for my pasture farming. I'd plant it first to a cover crop of some sort until I could deal with it properly.

But after I had the Farm Service Agency examine it to advise me, the report was shocking. Under Fred's stewardship, topsoil had been washing off the rolling field at the rate of twenty-two tons each year—from each acre. The government considered three tons an acre a tolerable soil loss. And the technician told me that Fred had drenched the dirt with weed-killing chemicals for so long that a cover crop might die. The cure was time and plenty of ground limestone, which would bind with the toxins and help neutralize them. I hadn't thought to have the field checked before we bought it. But the truth was, I hadn't cared; I'd just wanted it.

As Mike Guthrie put it when I told him of the damage to my proud addition to Mossy Dell, "He's corned that field to death."

"Where's Kathy?" Bailey asked, standing over our stove and stirring home fries. He was eight years older than me, but hyperactive and lean; in fact, he looked like he'd lost weight, from the way his wide leather belt creased the waist of his work pants. I hadn't seen him in two years, since we'd rendezvoused in South Carolina for a fishing expedition.

"She got up at six and went out to the farm," I said from our table, where I was drinking coffee.

"She went out there alone?" He raised his eyebrows above his horn-rimmed glasses; a bushy brown beard and full mustache hid his mouth. Steam rose from the cast-iron skillet. "Do you have any idea how unusual your wife is?"

"Sure. I married a horse. As soon as we get the kids fed we need to bring her a feed bag. And some coffee. Also diesel. Are you hungover?"

"What do you think?" Empty green beer bottles and a half-empty fifth of caramel-colored Wild Turkey bourbon sat on the counter. When lubricated the night before, Bailey had said he was surprised by how much I had bitten off. "I'm terrified for you," he'd said, and I'd suppressed a panicky feeling.

The day before, on Saturday, Bailey had traveled from North Carolina to help plant my future pasture and bulwark against development, which baked in the sun that May and struggled to clothe itself in healing weeds. Limestone would have to wait, at least until the ground was firm, but I'd asked him and Kathy to help me plant it. Kathy considered farm tasks optional in the face of two other looming urgencies: selling our house—though we had prospective buyers, another university couple—and getting started on Fred's dilapidated house. I'd prevailed, probably because it went against her upbringing to see dirt bared to the sky in the growing season. And we couldn't begin work on Fred's anyway until closing in six weeks. Bailey was curious, I think, after all my complaining. But he also wanted to help me out of a jam.

On Friday I had felt awed by the field's size—twenty barren, wide-open, rolling acres—when I'd driven onto it atop my shiny used John Deere. The tractor was compact, sized for tending Mossy Dell's irregular, hilly pastures. Though bigger than Dad's bantam machine I'd ridden in Indiana, it shrank beneath me on the expanse of soil. It felt like a toy, as did the small disk harrow I'd bought at Farm and Fleet to prepare a seedbed. For nine hours I had run the tractor along the old rows, north to south; the shiny, rolling disks toppled and shredded the brittle stubble. But Fred had planted the field in rows running in the same direction for so long that it had small ridges and swales, a corduroy effect that couldn't be seen but that jarred the base of my spine when I turned and crossed the ghost rows. I envisioned bouncing across the ground in future years when I mowed or made hay.

So Saturday, while I watched the kids and fetched Bailey at the airport, Kathy had taken over and disked in the opposite direction, trying to level the field. She'd learned tractor work starting at age eleven, disking, cultivating, and eventually plowing the fields of the Krendl farm in northwestern Ohio. Karl and Mary Krendl grew wheat, soybeans, and field corn, and they raised sheep for years—a horned ram once butted Karl and broke his leg—but they focused for income on growing thirty acres of sweet corn, tomatoes, and cantaloupes.

Karl built a white clapboard produce stand beside the barn. When sweet corn was ripe, cars crunched into the gravel driveway off Route 66 at the sign for Krendl Produce Market. The five beaming Krendl girls stuffed paper sacks with ears of Silver Queen: sixty cents a dozen, and they threw in extra. The income was intended to support their college educations, and the customers knew it. One regular always told them, "I'll buy your books for you."

From the first days of our courtship, I saw Kathy's pride in her driven father, the son of a man who immigrated from Austria and a woman who came from Germany. One day at Ohio State when she was making room for me in her Toyota, she opened its trunk to reveal one of Karl's pale blue post-office uniform shirts, neatly folded, that she'd kept. We were going to buy the tombstone he'd requested: an uncut boulder, inscribed only with Spencerville's zip code. "Imagine," he'd told her, "when some later civilization digs it up and tries to figure out what it means."

For their college savings, Karl gave his son and each daughter, in turn, responsibility for the crops and for filling canning-factory contracts for tomatoes. The kids marshaled the labor of their siblings to plant, till, and harvest. The harder his children worked and the dirtier they got, the more Karl loved it. They'd learn that life depends on the most basic things—seeds and soil and rainfall—but they would change the world with their minds.

Kathy, proud and bookish, didn't think it was fair that she did the tractor work for everyone, but Karl was unsympathetic. "You want to go to college, don't you?" he said, making clear her province. She did enjoy her competence and seeing the freshly worked soil and tender seedlings. She also liked being able to think, daydream, and sing songs from Broadway musicals at the top of her lungs. She never looked back to admire her work, because

Karl had told her that the trick to staying on course in the endless fields was to aim at a landmark on the horizon.

Right before her senior picture, she took sewing scissors and sawed off her long brown hair—it was hot on her neck in the fields. Her mother, mortified, begged the photographer to do something. He dabbed brown paint on the large photo Mary bought. The yearbook showed a girl wearing horn-rimmed eyeglasses and an uneven pageboy that looked like Joan of Arc's dented helmet. A layer of baby fat still softened her cheeks, but her composed smile, a young intellectual's nod to the world, wasn't warm enough to distract a viewer from the intense inquiry in her eyes.

Now, in the glory of late spring in Appalachia, she drove and disked for me. She was out there for her second day, my farmer's-daughter wife with a doctorate, gamely attacking the field's rippled ground.

Bailey strode across the dirt clods, carrying a bundle of stakes under one arm. I followed with a hammer. As Kathy disked the field's far side, we were dividing it into quadrants to apportion seed.

I had bought field peas, largely because legumes make their own nitrogen, but was worried that we'd run through the 1,400 pounds halfway through planting. Bailey took the lead, drawing on an experience from the summer of his sixteenth year when he'd measured tobacco fields, paid by the federal government to ensure that farmers weren't growing more than their quota. I pounded stakes where he said. Fifty-pound sacks waited in my truck.

Tomorrow would be Memorial Day, and blowsy white peony blooms glowed in yards shaded again by trees in leaf. It felt like high summer. We worked under a cloudless azure sky. Across Ridge Road on the hilltop, Fred's house stared down on us. In the pastures around it, meadow grasses already waved silvery green tassels of ripening seed. Even from the field I could see that Fred's lawn—ours, almost—needed mowing.

Although Bailey seemed confident in our work, he told me years later that he'd felt intimidated by the field's size, and unsure how my tiny seeds could possibly grow into plants to cover it. I was nominally in charge but fought to suppress disorienting waves of anxiety: I'd never dealt with this much bare dirt. I was responsible for this huge naked field, for any cloudburst that would cut rivulets in its face before my peas could protect it. A favorite phrase of our county agent's ran though my mind: "Be careful what you wish for. You just might get it." The weather was dry, and in breaking the field's crust, we had released the moisture from its lumpy clay face into the sky. I pictured Bailey's potatoes steaming in their skillet, and watched as Kathy made a pass and the dark freshly chopped soil behind her blanched to a pinkish hue.

Soon I saw her cut across the field, the disk raised, and park the tractor beside my truck along Snowden Road. We walked over as she dismounted and pulled one of my baseball caps from her head, dust rising from the hat as her brown hair spilled out. She chugged Gatorade from the cooler on my tailgate.

"That's it," she said, suppressing a gasp after gulping the liquid. "Two diskings." There was strain around her eyes.

"It looks good," I said. "Do you think it smoothed out the old rows?"

"Maybe a little. That disk doesn't cut very deep."

The miniature harrow was shy of 800 pounds, and I should've known its limitations. Using such a tool was like trying to flatten the ocean's swells with a motorboat. Heavy machinery had created this problem, and only heavy machinery could fix it. Kathy might have done it with one of her father's old tractors. Even his smallest, a blue Ford that she'd shown me once in her father's barn, "her tractor," was four times larger than ours.

In Mossy Dell's farmyard, Bailey and I hooked up Mike Guthrie's conical seed hopper to the John Deere's three-point hitch and then connected my cultipacker to the drawbar, at last using the implement I'd purchased before I had a tractor to pull it. Beside the field, we hoisted white paper sacks and poured fat green field peas into the red hopper. Kathy drove, seed flew onto the ground the disk had ruffled, and the roller obediently followed, firming the soil.

At last she'd sowed everything, almost a ton of seed invisible across the twenty acres. It was five-thirty, and we convoyed home to drink beer and grill hamburgers for the kids.

For ten days the sky stayed cloudless, and I worried about birds eating the seed. Finally a gentle rain, perfect for seed germination, fell on our thirsty acres of costly dirt.

CHAPTER EIGHT

To train animals, you've got to understand them, and yourself, first.

—my barber, Jim

ON A FINE DAY IN EARLY JUNE, CLAIRE AND I WERE DRIVING BACK TO Ohio over Maryland's western mountains. Fifteen ewe lambs stood behind us in the truck's bed, under a blue aluminum topper, a new $500 investment that transformed my pickup into an animal transport. The lambs were mostly white, some with black spots across the nose, and one was reddish brown.

"How about Cream?" Claire asked, trying out names based on colors and ice cream flavors.

"Sounds good," I said.

"Chocolate for the brown one."

"Naturally."

I worried about our climb of Maryland's Sideling Hill, 1,269 feet high and bracketed by runaway truck ramps; highway crews had cut a notch in the crest for the road, and from miles away I could see the defile ahead like a giant rifle sight in the rock. We made the climb easily and shot toward West Virginia.

The lambs baaed once in a while, as if suggesting we stop at the green fields flashing past, and flinched as tractor-trailer rigs swooshed by us. Looking in my rearview mirror and into the topper through its sliding window against the truck's cab, I noted the ability of sheep to compact: I could have stuffed in ten more. But I'd cleaned out Jo and Hank Simon's ewe lamb crop.

Loading the lambs was a fiasco. Rule number one in handling sheep is to crowd the critters into as small a space as possible, and I was now clear on this principle. Presumably Jo and Hank knew how to handle sheep but hadn't gotten suitable infrastructure established. Jo's lambs were free to flee from one end of the barn to another, which they did, forcing desperate tackles by Jo and Hank and a sturdy young neighbor who came over to help. I viewed and approved each struggling lamb before the neighbor lifted it and heaved it into my truck. The folly of this pretense at selection became apparent as most of the lambs stood under my blue topper: I had committed to buy fifteen lambs and Jo had only sixteen. The final two lambs were runty. I shot Jo a distressed look. *You expect me to pay you $100 for these?* She just stared back. When I mentioned getting only fourteen lambs, Claire protested that the little one with the poopy butt was *cute*. On the truck went the spindly creature. Claire named her Butter Brickle, I think; to me she was forever Runt.

As I drove, I thought about our other animal acquisition, currently in Kathy's care. Just before our trip I'd bought her, as an early birthday present, thirty guinea keets. I hadn't been able to resist when the Athens feed store started practically giving away a well-started batch of the pheasant-like fowl. Once a fixture on every southern farm, guineas were experiencing a surge of popularity because they eat ticks. Evidently they weren't popular in Appalachia, probably because people hadn't forgotten that guineas scream insanely at any disturbance—even a garden hose that looks to them like a snake—and their racket hurts the ears of any mammal. I had a soft spot for them because Mom's proliferating flocks were fixtures of the Georgia farm and the Florida nursery, where Kathy had grown fond of their comically alien ways.

I rigged up a pen for the quail-sized fledglings inside a fenced garden uphill from our house in town. I felt a tingle of worry, thinking about the guineas, because we'd have to move them before they matured, lest they blast neighbors with their alarm cries. Yet now, with the lambs and guineas adding to our ten surviving hens, the ones that hadn't make that fateful trip to Willy's, it felt like I was at last launched in actual farming.

At Mossy Dell, I drove the truck up the hill into the north pasture. It was a sweet early summer day, the field's craggy hickories and black locusts throwing heavy shade, and the sun-heated grasses rippling in a light breeze. Like anyone with a farm, I saw work. Already it was time to mow to keep the forage tender for the flock. We got the lambs inside Massey's shed by dropping the tailgate and shooing them out; they scrambled and slipped, their hooves spearing through the hay I'd laid in the truck's bed. Although I had my fancy German fence charger ready, this confinement in the shed was quarantine: I wanted to deworm them before releasing them onto our pastures.

Jo had said that she'd never drenched, which I believed, given her low level of management, the greater parasite resistance of Katahdins, her modest stocking rate, and Hank's cattle that vacuumed up sheep parasites. I knew I'd bought some parasites, though, and I enlisted Kathy's help in drenching our lambs the very next day according to a drastic Australian protocol for taking sheep to clean land that I'd read about on Sheep-L.

We marched uphill to Massey's shed through a sleepy, sunny Sunday afternoon. Kathy carried our bathroom scale, and I gripped the handles of a canvas tote that clinked with bottles of three different worm toxins. The lambs ran from us as I opened the gate, and I shooed them into a pen made from welded-wire fence panels along the shed's north side.

Kathy sat where I'd positioned her, on an overturned white plastic bucket, the scale at her feet, as I entered the pen and lunged at the lambs. They darted away like minnows and then reformed as a group in the far corner, looking at me. I hadn't made their pen small enough. I took another run and they eluded me again.

"Do you want me to come in and help you?" asked Kathy, dressed in jeans and running shoes and looking cute wearing my green baseball cap.

"No, stay there!"

Already I was yelling at her. Supposedly I knew what I was doing, but didn't, and felt frustrated. I used to hate it when I'd go boating with Dad as a kid and he'd get tense and yell orders. Now, angry with myself, I'd snapped at her.

"It looks like it's going to take two of us, Richard."

"How the hell are we going to get a lamb out for you to weigh it if we're both—" I stopped, realizing she was right. "Okay. After I catch one you untie that panel and help me drag it to the scale."

The first lamb I caught dragged me. "Kathy, help me, goddammit!" I yelled. I had grabbed the biggest lamb around the neck, and she was jerking me across the pen; then the lamb collapsed and I fell across her, pinning her with my chest, my knees on the hard-packed floor. Kathy ran over and put a bear hug on the creature, her arms around its middle. The lamb went limp as she hoisted its front feet off the ground.

We had our system. I caught a lamb as best I could, Kathy grabbed it around the shoulders, I untied a panel, and she dragged its butt onto the scale. This one weighed eighty pounds. I calculated the dosage for each drench, shot the dewormer down the lamb's throat with a plastic syringe, and herded it into an adjoining pen. The idea of the three different medicines was to hit the worms with an overwhelming onslaught and kill every last one. We'd have to do this again in only twenty-four hours, but the promise was freedom from drenching—forever. I hadn't dared mention this scenario to Mike Guthrie, fearing his scorn, and the veterans on Sheep-L had scoffed when the scheme was aired. Yet a shepherd had clean pastures only once.

We were soon hot and tired. The range in the size of our lambs was enormous: Runt weighed only thirty-five pounds and the biggest was eighty-five. By the end—almost three hours later, a ridiculous time expenditure for drenching fifteen lambs—we felt battered and bruised. But we'd gotten it done.

"Sorry I yelled at you," I said as we walked back to the farmyard.

"You've been doing that a lot lately."

"I didn't yell *once* when we planted the field."

"I was alone on the tractor the whole time."

That night in town we set our alarms for four-thirty in the morning, so we could drench again before we went to our office jobs. We drove to the farm in the dark, got there as the sky began to lighten over Lake Snowden, and wrestled sheep till past full sunup.

We made it back to the house in time to change clothes and race to work. That night, when I read up on proper deworming technique, I learned that it's easy to shoot the drench into the lungs accidentally and kill lambs. Especially if they were held upright as Kathy had helpfully positioned them for me.

Our lambs survived, though, and in two days I'd release them and be a grass farmer at last. The casualty was our bathroom scale, which never worked right after its use in Massey's shed.

The next weekend, across the road from Mossy Dell, Kathy and I visited the house we'd agreed to buy. The closing would be in a month, and it was finally vacant, Fred and Dolores having moved to their new home. Dolores had given Kathy a key. The weedy gravel driveway and the rank lawn made Fred's rough farmstead feel utterly forlorn. Earlier in June, Fred had held his auction. We hadn't attended, figuring it would be depressing to see curious strangers parking in the grass and swarming over our drab new home place, acquired, after all, only to protect Mossy Dell.

While Kathy tried to figure out what she could do with the musty house, I walked past Fred's three battered sheds, now empty save for litter. Two cracked concrete pads, strewn with nuts and bolts, marked where the galvanized corn cribs had stood. I entered the dark barn and saw that the fence charger was gone. Of course it was gone, auctioned or taken by Fred or his son. I'd have to buy one to bring sheep to the hilltop. I could get that model for $200 at the feed mill in Athens, half the price of my German charger, but its loss galled me. It had been the lone, pitiful extra promised with the place.

I saw tracks in the grass from the barn into the pasture and followed them to the field's far corner, where I found a tangle of bent pipes. Fred had piled the steel above a little pond and behind a thicket, which I surmised concealed an older dump site. As I returned to the house to report this infraction to Kathy, Fred and Dolores pulled into the driveway in

Fred's black truck. Friends or relatives had probably called them to report our visit. Unlike Mossy Dell, a secret valley world, the house on the hill was bared for inspection from a mile away.

When Fred saw me walking toward the house, his eyes darted past me to the tracks.

I decided to say something: "I would appreciate it if you would stop dumping things in the pasture."

"Oh, I can get rid of that stuff if you want," he said, as if he were offering a favor and hadn't violated his agreement.

"Just don't dump anything else."

"I need to use the bathroom," I heard Tom say to Maya. It was Tom's turn to ride Dream, the horse I'd bought for Claire from Maya and her husband, Bill, members of the grazing group. One day at a pasture walk, Bill had announced that they wanted to sell the gelding that was their daughter's childhood 4-H project. Now they were boarding Dream until we could get situated on our land.

"Do you need to pee or do you need to shit?" said Maya, a short stout woman wearing a green sweatshirt, its sleeves cut off, a red bandanna tied over her hair. As I walked up, I saw Tom hesitate. "Because if you just need to pee," she said, "you can go behind the barn."

"Pee, I guess."

"Come on, Bud," I said. "I'll go with you."

"We have a composting toilet," Maya explained. Just past her, Claire was glaring at me. She wasn't happy about more lessons, considering herself an expert after some instruction back in Bloomington.

"I'll ride again!" said Claire's friend Lucy, a blond sprite who lived in town with her divorced mother and who was always eager to go with us. She launched herself aboard Dream. Maya was holding a fuzzy yellow lead rope on the horse and began walking with Lucy in the saddle, holding the reins. Dream plodded along. Claire followed, looking at the ground, her chin tucked into her dark blue T-shirt.

From the start, she'd protested that Dream was a *pony*, and I couldn't convince her the stocky brown and white animal was, technically, a small horse. I'd bought him and his tack for $1,000, without dickering.

My labor on a corral for Maya was part of the arrangement between us for her lessons and for their boarding of the horse until we could take him to the hilltop. It was a round training pen, about sixty feet across, that Maya said would help motivate and discipline Dream, who'd been balky. One day in the barbershop, Jim had loaned me a videotape, *Round Pen Reasoning*, that I'd watched to understand Maya's refresher training. The method involved the trainer, standing in the corral, turning the horse this way and that by moving toward the animal or backing away as it circled.

Finally Maya had nagged Bill into sinking posts for the enclosure, uphill from the barn in the lee of a woods, and I was laying out its rough-cut poplar boards. A shadow crept toward me from the trees and I shivered in my short-sleeved shirt; it was cool for July, fall-like, a river of cold air from Canada having lifted a miasmic summer haze. I was wearing gloves, but my right hand hurt from a recent injury, from trying to tackle Runt. She was a problem, escaping daily from the dry, sunlit front corner of Massey's shed, where I'd confined the lambs until they finished pooping out parasites. Runt was small enough to wriggle through the bars of the shed's front gate, and always scrambled back inside when she saw me. The day before I was to turn out the flock to graze, she was out again, and this time ran into the shed's dank rear, an alley behind the lambs' makeshift quarters. I cornered her there in the dark, cattle-fouled compartment. As she scooted past me, I leaped to tackle her and missed, falling heavily onto the muddy floor. Something stung my palm—a rusty nail sticking out of a weathered board. As I got to my feet, I saw Runt slip through the gate and rejoin her flockmates.

I drove myself to the hospital emergency room in Athens, contemplating the ideal conditions for infection under which my wound had occurred. When I explained that I'd been hurt in a wet, manure-encrusted barn, the staff gave me, in quick succession, a fresh tetanus shot in the arm and a Betadine soak for my hand. I felt proud of my first injury. Then a nurse scrubbed the hole with a sponge, and her treatment hurt more than the nail had. My prideful amusement vanished completely when a doctor probed the puncture with a blunt wooden stick.

"Dad! Dream bucked Tom off!" Claire yelled uphill toward me. I dropped a board and ran toward her. She held her arms stiffly out in dramatic beseeching, her gesture incongruous above her baggy khaki shorts.

"Was he hurt?" I asked.

"No, I don't think so. He landed on his butt."

Farther downhill, Maya, Lucy, and Tom stood looking at Dream, who grazed nearby, reins knotted atop his withers. Maya held the lead rope.

"Are you okay, Tom?" I asked, my hand on his back.

"Yes."

"Dream needs more control," Maya said. "We have to get that round pen done."

"I'll come back to finish getting the boards on. I didn't know Dream bucked."

"He used to sometimes," Maya said. "He threw our daughter when she was twelve"—Claire's age!—"and broke her collarbone."

Driving back to town, Tom in the passenger seat, Claire and Lucy in the truck's extended cab, I carried on a frantic inner monologue. For all my worries about horses, I hadn't checked this one out. Why hadn't I asked Jim for help getting Claire a horse? Or at least had him evaluate Dream? Because I thought I knew the owners? Because it was convenient? Because he was *pretty*? He probably hadn't been ridden in years. Dad said horses that aren't ridden get lazy and treacherous—"barn sour," he called it.

"Dream is dumb!" Claire blurted behind me, her voice loud in my ears. "And I *hate* Maya. She's mean. And her house *stinks*. When I used her bathroom, the fumes burned my nose."

"She seems nice to me," I lied, "taking time to give you lessons—"

"I don't *need* any more lessons. And she's only nice when *you're* around. You're always talking to Bill. You don't know. She orders me around like I'm a little kid."

"You sound like a little kid. Like a spoiled brat. A lot of girls would *love* that horse. And they'd *thank* a Daddy who takes them riding."

"You shouldn't have bought Dream!" she yelled. "You should have gotten a different horse. A *horse*. Not a *pony*. You got Dream so *you* could hang out with Bill!"

Dream *was* a terrible mistake, this incident the final straw. Yet I bristled at Claire's anger, her ungratefulness. I'd regret not admitting my own feelings to her. But in the moment I couldn't; pride stopped me. I couldn't see then that her rage was really about *everything*: the move, losing her Bloomington friends, clueless me. I gripped the steering wheel and shook my head self-righteously. I stared out the windshield at the brushy countryside. Stubbornly I let the angry silence linger and solidify. I hoped her ears rang with her own harsh speech.

"I'm sorry," Claire said.

From her tone I knew her frown was gone, her face open and vulnerable as she looked at the back of my head. I waited a beat too long to accept her apology, to talk. As the moment passed I could feel us sink, estranged, into bleak inner worlds. In the silence, I steered the truck through a curve. Even Tom, naturally quiet, now seemed to radiate lonely apartness.

I remembered Lucy and glanced in my rearview mirror. She and Claire sat opposite each other on the bench seat. Lucy was turned, looking out the window, her face as veiled as Claire's.

I walked into Ernie & Jim's Barbershop, clutching issues of the *Stockman Grassfarmer* for Jim and his horse-training videotape. Jim, wearing a white Western shirt with snap buttons, lounged in his barber chair in the empty shop, roosting in the window wall's golden light like an old-time porch sitter, doing nothing with palpable enjoyment.

I knew he was dreaming about his farm. Around a bend from his father's idle land, Jim trained horses; he often spent weekends at auctions down in Kentucky, buying and selling. Despite Ernie's traumatic farming experience, Jim was living his own rural dream.

He'd warmed to my proselytizing about grass farming. He liked the idea of his horses grazing, rustling the range like livestock did out west, instead of standing around all day in a stable waiting for him to feed them. Jim wasn't the average horse trader. After his boyhood spent riding the roads and countryside around our farm, he'd participated for several years in the annual roundup of wild horses at Chincoteague Island, Virginia; he'd admired the old-time cowboys he met there and loved their campfire tales. He'd since become a serious student of how horses saw the world. People called him a "horse whisperer," a trendy term he seemed to dislike, perhaps for its mystical pretense or its whiff of hucksterism. Yet horse whispering was shorthand for everything people like me didn't understand about how humans, fragile physically in the natural world, could meld with and bend to their will big, strong, mercurial animals.

Jim spun me toward the mirror and combed my fringe. I felt agitated and wanted answers about Dream, but was keenly aware I'd disrespected Jim by getting that horse in the first place. Still puzzled as to why Diana's cow Charlene had tried to kick me, now I had a horse behavioral problem, and soon I'd be handling a grown ram.

"Hey," I said. "I found an old photo of a pony in the cabin."

"Kenneth and Mabel were Tennessee Walker people," he said. "They raised a few Shetland ponies in the '50s and '60s for work in the coal mines. That was probably their stallion."

"Is that one of Kenneth's horses?" I asked, looking at the barbershop's only ornament, a framed photograph hanging in the mirror's upper right corner, in front of Ernie's chair. Fading badly and blasted white at the edges, it showed a teenaged Jim with tousled hair beside a horse.

"No," Jim said. "That's a quarter horse. That gelding was my first good horse."

Dad had taught me to mistrust horses. A talented rider who my mother said had "learned to ride before he could walk," he worked cattle from horseback in California and had told me about horses' treachery, their kicks and bites, and about one that always tried to dislodge him by running under tree limbs. He'd forbidden riding lessons for his kids. So I would've felt safer saddling up Charlene, who at least wouldn't be expected to enjoy my presence. But I acknowledged horses' physical beauty. And Jim's gelding, poised in the photo at the fulcrum between power and grace, looked magnificent. Young Jim stared at the camera, and the horse faced him, turning its muscular rump to the photographer; the horse studied Jim with ears raised, its head lowered from an arched neck.

Jim's quarter horse was named Eddie Bueno—"Good Eddie"—grandson of a legendary stallion named Poco Bueno, "Little Good." I later googled Poco Bueno and confirmed that he was a famous sire; when the horse died at the age of twenty-five, to honor him he was buried in a standing position on his owner's Texas ranch.

"Eddie was worth $2,500, but I paid only $400, because he bucked," Jim said. "His owner had screwed him up. Once a horse learns how easy it is to throw humans, it's hard to resist."

"Could you straighten him out?" I asked, my thoughts returning to Dream.

"I did, but it took time. Dad was running about a hundred head of Angus, and I found out Eddie wouldn't buck when he was in front of a cow. He wanted to be a cow pony."

"What happened to him?"

"About five years after that picture," Jim said, "somebody shot him."

"No!"

Jim lit a Marlboro and stared out the window. "It was deer season. This was the only part of the state that had deer then. The back roads were full of strangers cruising with gun barrels sticking out of their cars. Maybe it was a stray slug, or somebody just got bored and shot him."

I looked up at the photo and saw the day, Ernie ushering the photographer out to the warm field where Jim and Eddie Bueno waited. Tough-guy Ernie's desire to commemorate his son's passion touched me. There had been many horses in Jim's life since the gelding, but surely no other photograph taken by a professional hired to mark—and inadvertently to memorialize—a moment in time.

"That gelding I bought is stubborn," I began. "And he's worse with Tom. He hasn't been ridden in a while and doesn't seem to want to be."

"Makes sense," Jim said.

"The video helped," I said. "I guess I still don't know how to read him."

"It's about body language, the human's and the animal's," Jim said. "It's a silent language." He snipped at my hair, contemplative. "It's all there, in that book," he said, gesturing with his comb toward the shop's magazine-covered table. The thick treatise was nestled among journals devoted to humans with large mammary glands. Jim pulled it from under a copy of *Playboy* and handed it to me: *Communication and Expression in Hoofed Mammals*, by Fritz R. Walther.

I couldn't believe that in a redneck barbershop the size of Diana's milking parlor, a German scientist's magnum opus lay in my hands. I opened it to Walther's drawing of two zebras interacting, their long ears and narrow shoulders set at telling angles.

The bell on the shop's door jingled and Ernie entered, carrying a Styrofoam cup of coffee. "Hi, Richard," he said. "I guess we're going to be fence-line neighbors now that you're adding Fred Paine's place to Lost Valley."

"Oh, yeah, we're negotiating to buy Fred's," I said—as if I'd been absent-minded, as if we hadn't reached agreement—aware of an edge in Ernie's voice. I was impressed by his grapevine and embarrassed that I hadn't mentioned it before. Ernie might have seen us out planting the cornfield. His little house was just around the curve from Fred's hilltop, on the same side of Ridge Road. Because we drove to the farm from the opposite direction, I could forget that Ernie would be one of our closest neighbors.

"My land goes all the way to his northern border," Ernie said. "Fred's been telling everyone he sold out to some dean."

I'd never mentioned Kathy's job to Ernie or Jim. So even Fred knew her title—of course he did. "Dean" sounded important, even if I had a hard time myself explaining what she did as an administrator except coordinate academic programs and raise money.

"Sorry about those junked cars," Ernie added. "I'll get them out of there."

"I didn't really notice them," I lied.

"I just hauled them over there by his house to piss him off. That sonofabitch has tried to screw me and every other person up and down Ridge Road. You watch yourself."

"Thanks, I will."

"He might try stuff with you," Ernie said, glancing at me, draped in a plasticized black smock in Jim's chair, "but he won't mess with your wife, that's for sure. I hear she's plenty tough."

This comment hung in the air. In Ernie's assessment I was a "nice" guy, harmless enough—that is, weak. How *did* I come across? Not like a farmer, probably, with my slight build, my eyes puffy from allergies behind stylish glasses, my office shirt. It sounded like a professor getting his hair cut had given Ernie an earful about Ohio University's newest dean. Yet Ernie spoke with admiration, so his informant probably was our acquaintance who owned a stable, a senior professor who'd welcomed Kathy's ideas and energy. Claire had taken some lessons at his place before I bought Dream.

"She's making changes," I said. "Kathy moves fast. She'll have done six things before anyone starts second-guessing the first. They had a saying back in Indiana about people like her: She goes at it like a dog killing chickens."

Ernie, sitting in his chair gripping his coffee, grunted appreciatively and said, "Fred has finally met his match."

"Speaking of Kathy," I said, lifting my head to address Jim behind me, "she heard from John Baker"—the stable owner—"that one of her new professors got kicked in the face by his mare a couple of weeks ago. I guess he took a bucket of grain away from her and she whipped around and nailed him, caved in his cheekbone. He's got to have surgery. Kathy's worried because we're now in the horse business ourselves."

"She treated him like another horse," Jim said.

"Probably not like the lead stallion, either," I ventured.

"No."

"How do you avoid that, with all your horses? Some must act up."

"I'm the dominant horse in the herd," he said.

"What if one gives you trouble anyway?"

"I'm no rougher to a horse than a horse is to a horse. Sometimes that's plenty rough. But you only have to do it once."

Evidently, getting physical was a less-publicized backup tool in the kit of the horse whisperers. Jim swept hair off my poncho with a whisk broom. "In any relationship," he said, summing up his philosophy and apparently his irascible father's, "one is the hammer and one is the nail."

Over at Mossy Dell, our flock grazed. When I visited, the lambs looked up at me with placid curiosity as hanks of grass disappeared into their mouths. It was strange to be regarded by an animal in this way, as a diversion in their timeless day.

I didn't see it happen, but they'd trained themselves to the electric fence and respected its bite. The German charger sat clicking in Massey's shed, sending bursts of electricity into an insulated cable that energized three wires I'd strung down the middle of the five-acre field on white fiberglass stakes. The electrified division fence formed one edge of the sheep's paddock and electrified the other sides, movable Electronet mesh that enclosed them in what the fence supplier called a "pain-filled grid." Down one side of the field and up the other I moved them to fresh grass in their enclosures: a "racetrack" pattern, in graziers' slang: around and around the field the animals went in their Electronet, pivoting around the fixed fence.

When Doty and Jack touched the Electronet, it sent them yowling to the farmyard. After that, Jack wouldn't go near it, and Doty would cower under my truck when I'd head into the pasture. She'd gotten hit by the charger's voltage twice, once going under the Electronet and again after she panicked and tried to run back through the fence to get to me for protection from the unseen source of her misery. The charger had been on its highest setting, so she experienced as much as 9,000 volts. Modern chargers are relatively safe because they move electricity in short pulses instead of in a sustained flow.

One sweltering afternoon, perspiration rolled off my brow as I knelt in the grass and gingerly tied a roll of Electronet to a post of the electrified fence with string. I was being careful because I didn't want to bother walking back to Massey's shed to turn off the charger. Heat rose from the grass and flushed my face. As I tied off the string, my sweaty forearms drifted upward. I didn't feel my skin touch the charged wire; I heard howling— chest-deep, feral: my own cries in my ears—as almost 4,000 volts, the charger's low setting, surged through my well-grounded body. The pain was otherworldly, peculiar to electricity; it jolted me at some primitive level, my reaction too complete and too quick to form the thought *Pain!* One moment I was working, and then my body jerked and I hollered and reeled backward. The episode was unforgettable, yet the only evidence was a small red mark on my right forearm.

The dogs and I had inadvertently provided reassuring proof that the charger worked and would protect the flock from coyotes. The sheep, less conductive on their flinty hooves, just flinched and jumped back if they brushed against the fence.

"Oh, sheep!" I'd shout to summon the flock to fresh grass. This call, cast into the unpopulated landscape, had a plaintive quality that began to bother me. It also sounded like I was yelling a curse, a frustrated cry over our unsettled situation, into the indifferent hills. I asked an experienced shepherd on Sheep-L about this, and she told me the call—"Ovee!"—that shepherds have used for centuries. Derived from the sheep's Latin name, *Ovis aries*, this cry carried far, like the cattle herder's traditional "Come, boss!"

I'd wade into the pasture every afternoon and wail, "Ohhhhh-veeeee!" Somehow it sounded right, musical even, despite my poor singing voice. It filled my mouth, throat, and chest, conveying an urgency the flock understood. The sheep jerked up their heads from the grass and followed me as if I knew what I was doing.

A tame sheep emerged that summer in Mossy Dell's shady, fertile pastures. With the exception of one ewe, Cream, I couldn't remember their ice-cream names, so I called her Red. Although her body was white, she had a patch of reddish hairs on her neck.

Red always left the flock to greet me, to be petted, and to seek an extra helping of the oats I was using to try to tame the flock. Why one sheep out of fifteen had decided to make friends with me—the scary human, the pushy herdsman—intrigued me. I treated them the same, tried to tell them apart, to get to know them, but only Red was friendly. Nothing special to look at, small and fine-boned, she got extra attention by being tame. In this way, Red became a leader, the ewe who helped me take the flock to fresh grass, even if we had to cross the farmyard, past appealing grazing opportunities. The wilder ewes, like difficult children, had their reasons, but they didn't endear themselves to the shepherd. And their behavior probably made it more likely that their suspicions would come true. Red trusted me, and influenced me to reward that trust.

Compared with wild animals, all the lambs were docile. Yet only Red enthusiastically chose to affirm an ancient bond. The ritual of domestication, a milestone in human history, was being reenacted on our Appalachian farm. In the apparent absence of meaningful selection by Jo, nature had thrown out a range of temperaments—from deeply suspicious to ridiculously trusting—to see what was needed for survival on our farm.

Mike Guthrie was a fan of stockman Burt Smith, and I studied Smith's book *Moving 'Em*, which emphasizes that animal behavior is both genetic and learned, the same as in humans. Smith gives an example of learned behavior:

> Some time ago a hunter was hired to kill off a herd of about 40 elephants, but he did so very slowly and they learned to fear him. When the herd was down to about 18 head it changed its habits and became nocturnal, and also attacked humans if followed into bush. The herd was granted a reserve in their area, and no hunting permitted. Several generations later, the elephants still have the same behavior towards humans, even though none of the existing population has been hunted. Culture is not the sole invention of humans.

Our flock's culture was one of fixed wariness toward people. It occurred to me that Red had been one of Jo's bottle babies, and if so, she'd been tamed in infancy. She'd learned. Yet surely some of our other ewes had been given bottles too. Red was inclined to be tame. The puzzle of her behavior, her genetic inclinations interacting with the environment—*me*, in this case—fascinated her flockmates, who hung back, amazed.

Intrigued by Jim's copy of *Communication and Expression in Hoofed Mammals*, I checked it out from the university library and learned that Walther ultimately held a rather poetic view of human-animal relationships. Personifying animals—the alleged sin of anthropomorphizing—is actually the natural basis of humans' relationship with them, he says. And animals operate exactly the same way toward humans. This tendency deserves a monument, the scientist believed, even though human-animal interaction stems from the "error" of both sides treating the other as if they were the same species. "In particular," he writes, "'man's greatest biological experiment,' the domestication of large mammals . . . at a time when the human race was still in a rather primitive state of technical development, would have been impossible without man's anthropomorphizing these animals, and the animals' zoomorphizing man."

So, sweet little Red—and mean Charlene—had zoomorphized me. Red was treating me like a dominant sheep—a ram, from her flirtatious behavior—and Charlene must've viewed me like a submissive cow, with amused contempt for my cautious milking routine. My shouting after her kick probably sounded like a calf's laughable bawling.

As for Dream, like Charlene he was unimpressed with some humans' authority. I doubted Maya's ability to change his mind. I certainly couldn't do it. And even if I hired Jim to get physical with him, would Dream still defy—and possibly hurt—Claire and Tom? I'd have to put our horse visits on hold until I decided what to do. Chewing off my fingernails hadn't provided any answers. I felt like a man stuck in quicksand in the old Tarzan movies I'd watched as a boy: I could talk to bystanders, even catch my breath as I slowly sank, but I couldn't escape; desperate thrashing only took me down faster. I was making too many big decisions too fast, and humbling new problems resulted. Of course an underlying problem I couldn't see then was my pride. I could have simply given the horse back and taken a loss, but was too ashamed of my mistake and, weirdly, embarrassed for Dream's sellers.

Meanwhile, the scraggly cornfield, beside the road where everyone passing could see it, advertised what my withdrawal of Fred's chemical life support had wrought. The peas had germinated, but the stand was thin due to herbicide carryover, low fertility, or absent topsoil. As the summer lengthened, it appeared I'd wasted our time and $532 in seed. Parts of the eroded field couldn't grow decent ragweed on the sticky clay subsoil. I tried to comfort myself with the thought that we'd at least done something constructive to begin its slow healing.

And Red mooned up at me every afternoon with friendly, trusting eyes.

She knew that the fence caused pain and that I brought caresses and treats. She could communicate; she could think. And she'd reached a conclusion different from her sisters'. Red was an individual, which is to say a mystery. And a delight. Her antics helped offset Kathy's and my feeling of oppression as we contemplated what we'd bought. The closing was in two weeks. We steeled ourselves for a high-summer overhaul of our new house.

CHAPTER NINE

In my mind's eye the view from that brick house on the hill is in
some future summer, of a pasture where cows swish their tails and
sheep lie down sleepily in the heat.

—e-mail to Bailey

THE JULY EVENING BEFORE OUR SCHEDULED CLOSING, FRED TELE-
phoned and told me that a thief had come in the night and stolen the two
huge black lion statues that guarded his driveway. "I've reported it to the
sheriff," he said. I was relieved by this obvious lie. How could I have gotten rid of those pretentious eyesores? "Thanks for letting me know," I said,
and got off the phone. Interacting with him in person again concerned me
because by then it was hard just looking at his long bloodhound's face and
hooded eyes.

In a neighboring town the next afternoon, we sat in the airless conference room of Farm Credit Services and signed papers. Fred was smiling,
smug—the rat that ate the grain—and Dolores, dressed in a white pantsuit,
appeared tense. The banker, an earnest man who wore thick black-framed
glasses, had worked with Fred for years. He'd agreed to finance our purchase
as a working farm at lower interest than the going residential rate, as a favor
to Fred and because I'd promised that we intended to farm. At the end he

handed us a couple of stiff nylon baseball caps emblazoned with the bank's logo, and I shook Fred's big hand, surprisingly soft.

Now, in mid-1998, only two years since our arrival in Athens, we owned a brick house on top of a hill at the intersection of Ridge and Snowden roads with a huge barn, some tired sheds, sixteen acres of pasture, and a little pond. Across the road we owned Mossy Dell, enlarged by a fringe of woods and by a big, weedy, eroded field out front.

Even as I put our van in reverse in the bank's lot, Kathy was punching a preprogrammed button on her cell phone. The young builder we'd hired, Bruce Bledsoe, was waiting at the site, he and his crew poised to tear into the house with hammers and crowbars. At the closing we'd been acutely aware that we had just sixty days before vacating our own house, which had sold immediately. Bruce had said he could rip out the paneling, replace it with sheetrock, and remodel the kitchen and bathrooms in the short window we had. It was an aggressive plan, but if all went well we could just make it.

I had to admit that Fred had come through, too. The shifty guy had greased the skids with his banker and saved us money.

Within days the calls started coming. The county wouldn't let the bank's lawyer record the deed, because of Fred's previous run-ins with zoning officials. In Athens County, we learned, no more than five pieces could be sold off a single property without formal development—amenities such as roads and septic systems—and each parcel had to be at least five acres. This "five-split" rule and the acreage minimum were intended to prevent a hodgepodge of housing in rural areas, and to ensure that each homestead had enough land for a leach field. Fred already had broken his property five times, years before, in selling off the strip along one side of Snowden Road, across from the cornfield. And two of the lots were under five acres, violating the minimum-acreage edict. Although he'd sold us the cornfield and the nucleus of his homestead, he was retaining a crop field south of the house, thus indulging in yet another division of the property.

County officials had been waiting to lower the hammer on Fred, and we were caught between. We could sue him, or sue the bank for proceeding with the closing without checking on its legality. We didn't want a fight, though, because nobody could win. And we knew that neighborly relations must be preserved at almost any cost. The alternative is endless acrimony— your mailbox bludgeoned, your fences cut, your dog poisoned, your ram shot during deer season.

Besides, we had to resolve this amicably because we were in an impossible negotiating position. What had been a house was now a demolition site. Upon removing a sagging portion of the low ceiling, Bruce had discovered a dense blanket of black mold that covered every surface in the attic, the result of Fred's failing to install roof ventilation, a structural necessity and vital here, given the wet basement. He also found termite damage in wooden headers over doorways, and gray yellow-jacket nests packed into wall cavities.

We'd had the house inspected, but evidently it had been a quick and dirty job; anyway, we'd thought we knew its problems, and we planned to remodel anyway. When Bruce, appalled by the house, had urged us to let him gut it, to rip off the roof and take the walls down to bare studs, we'd agreed. Within days, a mammoth green construction dumpster in the driveway overflowed with debris. Fred's faded house was now our active mess.

"Strange place," Bailey replied to my e-mail. "That 'five-split' shit is bizarre. They don't even have building codes there and try to enforce zoning like that? I agree you shouldn't sue, but where was your lawyer in all this?"

We hadn't hired a lawyer. We'd never used attorneys for our real estate transactions in Bloomington; people there didn't, and they didn't seem to in Athens, either. Now I hired one, a woman who went with me to see the county planner. He turned out to be a genial fellow, another ex-hippie homesteader from the looks of his shaggy blond hair, worn Levis, and Nike running shoes. I pointed out that we weren't covert developers in cahoots with Fred to sell plots, but innocent farm buyers—*foolish buyers*, I realized, too late—who were left holding the bag. He listened and chatted briefly with our lawyer. He said he'd look at the property and talk with county commissioners.

Meanwhile, we couldn't tell Bruce to stop—it was too late, with the house now unusable and unsalable. By the end of the following week the county approved the hilltop and the cornfield across from it as a "minor subdivision." This technicality allowed the deed to be recorded, and spared all concerned further discussions about Fred's history and our intent.

Fred had achieved his objective—to sell his decaying house—and though he hadn't wanted to part with land, throwing in the cornfield had made his larger goal possible because of my salivating desire for more acreage. He'd known he could bluff or bulldoze anyone who got in his way. Zoning regulations weren't his concern, and he knew they were unenforceable.

What we had to show for my dreams and our money was abused land and a wrecked house.

I couldn't believe what I'd gotten us into, and yet now there was no turning back.

I helped with the deconstruction as I was able. Even accounting for the fact that I was fifteen years older than Bruce and his two sunburned, tattooed buddies, I was impressed with how they endured the daily physical demands of construction work. All day, every day, on their feet—lifting, bending, stretching, pulling, carrying, climbing, pounding. They were small-town boys, all ex-military—Bruce and one of his workers were Marines, the other an Air Force veteran—and they engaged in a warfare of sorts against material objects, the hot weather, and human inertia. They waded through rubble, the dumpster overflowed with trash, and smoke drifted from burn piles as if from an artillery barrage.

Within six weeks, Bruce spent our nest egg of $70,000 from the sale of our house in Indiana. We'd planned for that money to cover the entire costs of renovation. But at the start, unaware of the problems we'd face, we hadn't signed a standard fixed-price construction contract with Bruce, who'd said he didn't know how to estimate such an undefined overhaul; we'd taken the "time and materials" route, paying him an hourly rate. Then we'd agreed to destruction.

Kathy and I arrived at the scene one evening to find that the roof and the house's interior partitions were gone. Just three exterior brick walls stood under a blue summer sky. Leaving Kathy pacing the plywood subfloor, trying to figure out a new floor plan, I wandered into the hilltop's pastures.

Fred had made the twisted pipes disappear as he'd promised, though I would always wonder if they'd ended up in a more concealed location. Maybe at the bottom of the pond.

I towed my new trailer, stacked with our oak boards, to a carpenter's farmstead. He would mill the rough-cut planks into smooth baseboards and door and window trim. The boards had air-dried for more than a year since the three-legged oak had crashed to the ground. Then Kathy and I had wrestled the wood from the corncrib and barn at Mossy Dell onto the trailer and hauled it to a logger with a kiln. Now I drove through a lonely valley southeast of Athens, near the crossroads of Shade, to see the carpenter our builder had directed me to.

He ushered me into his house, a shipshape white farmhouse that was clad in real wood, not vinyl. The interior was finished in loving detail with cherry, walnut, oak, and maple; I felt as if I'd stepped into a museum of early American woodcraft. Bookcases, cabinets, and window seats gave off a buttery glow and displayed the natural beauty of various colors and grains.

He showed me different styles of trim. I could recognize the distinctive shapes and decorative patterns as he pointed them out, and could appreciate the tight joints and carefully fitted corners. But I'd grown up in a Florida beach house with scant built-in decoration, and had no vocabulary for this world of hand-crafted capitals and blocks, of beaded sideboards, capped baseboards, and fluted crown molding. Worse, I had no taste—it all looked good to me. Subtle differences were too subtle for my eye, which was accustomed to our Indiana house's simple picture-frame window trim: pine, and painted white at that. This region embodied a knowledge about wood and a reverence for it that was unique in my experience. Appalachian Ohio had engendered craftsmen who could construct a house, its cabinets and furniture, from the surrounding forest.

"You might like this, a local style," he said, leading me into his dining room. I readily agreed: the wide trim artfully fitted around his doors and windows looked fine.

I unhooked my trailer and drove away, thinking about what I'd just seen. I hadn't noticed there was a native trim design, of course, and was grateful for the carpenter's guidance, even as I was embarrassed to discover the depth of my ignorance. I wondered which was the truly local structure: Fred's cracker box built from generic blueprints, or the carpenter's glowing jewel box. Well, both, I thought. Fred, yet another refugee from West Virginia, was native to Appalachia if not to Athens County, and didn't his pragmatic house reflect *him*? Like our decaying log cabin at Mossy Dell, his moldy house hadn't been built for the ages and had lived its life. I found Fred's stinginess offensive, and in an era of scarce resources could openly condemn it. He and Dolores moved because what he'd built was falling apart—yet he now seemed almost enviably pragmatic. He'd found a buyer for his neglected post pounder and for his dying house. We were the ones paying repair bills and tearing down dwellings left and right.

The craftsman had preserved in his valley the highest expression of the second wave of settlement, a gracious frame farmhouse trimmed in burnished hardwoods. That's what I chose to define as local, what I wanted, the heritage our young builder steered me toward. Local culture is small and vulnerable, endangered by undue poverty, upheaval, and affluence. People must always recapture, relearn, reacquire it. Maybe come to define it. Our moldings and baseboards cut from Mossy Dell's white oak would speak in the vernacular, though here, as elsewhere, the local had been pushed to the margins and gone underground, lost both to inhabitants of utilitarian trailers and to the middle class in their trendy mansions. We'd blundered into a parallel counterculture. I could faintly hear a lost tongue singing from a vanquished woods culture.

After the carpenter delivered the trim, we sanded and stained it under Bruce's supervision. I couldn't believe how many times we'd handled that oak, from logs in Mossy Dell's wet farmyard to the hundreds of pieces of planed, sanded, stained, kiln-dried sticks spread across our gravel driveway on the parched hilltop.

One Monday our banker appeared onsite. I was across the road, moving the sheep to fresh grass, and Kathy was talking to Bruce about hanging the kitchen cabinets, discontinued display units she'd gotten at a bargain price. Kathy and I had had a tense weekend, enclosing a horse stall in the barn for our Ameraucana hens and the guinea flock, which I'd moved out Sunday night from town. The guineas were protesting, and their insane ruckus carried all the way to the house.

The banker walked around, looked up, turned in circles. "You've torn down a perfectly good house," he told Kathy. "You've put the bank's investment at risk."

His alarm was understandable, I now see. We hadn't asked him to finance the massive reconstruction, only the basic purchase. Kathy later told me she felt a tightening in her stomach at his words. Had Fred, embarrassed by our publicly gutting his home—and angry with himself for failing to squeeze more money out of us—sicced the bank on us?

"Fred let the place go a little," the banker admitted. "But the house was only twenty-three years old." Kathy took his point: Fred was local and we were crazy—outsiders too good for a decent house.

His letter came the following week. He'd converted our thirty-year fixed mortgage to construction loans, at a much higher, variable interest rate. He also demanded that we complete the entire project, including finishing upstairs accommodations permitted by the new, soaring Cape Cod roof. Kathy called him and argued hotly. "The house was almost uninhabitable," she said. "What we've already rebuilt is of greater value. Your money is more than safe." He compromised by giving us a second construction loan and returned to the hilltop with a form, which we signed with Bruce, certifying that all workers had been paid. This absolved the bank of any further responsibility and let us off the hook on finishing the upstairs.

In paying the interest on the construction loans and in financing so much of the work ourselves, we charged four credit cards to their limits. We borrowed against our pensions, sold the kids' stock, and spent their college savings.

One of Kathy's colleagues loaned us her house when Kathy mentioned that she'd have to find us an interim rental until Bruce got the hilltop house finished. "It's not a big deal," Bella had said. "I have my research in Washington through September, and I can stay with friends afterward."

Of course it was a big deal. We'd never felt so vulnerable. Exhausted, broke, scared, we knew another misstep could plummet us into bankruptcy and public shame. And Bella's house was unique: bought after her divorce, it was smack dab in the middle of campus, a social and intellectual salon for the university and community. Bella hosted legendary parties there. Tucked in a grove of white pines beside the music school, her house featured dark cork floors, orange shag throw rugs, and swag lamps dangling from shiny golden chains. It cried out for jazz music on a phonograph and the clink of martini glasses.

Within two weeks, with our dogs and cats confined to makeshift kennels in a shed at the new place, we were living in this gracious world, this alternate dream, Bella's pad. When her windows were open, we lounged in canvas sling chairs or sprawled on a giant beanbag, listening to students practicing their saxophone solos. Kathy said she felt like she was nestled in loving arms. For Claire, who'd be starting Athens Middle School, with all the anxieties of adolescence, the location was perfect, just a three-block walk; friends could come and go freely. Tom wasn't old enough yet to walk to East Elementary alone, but he enjoyed his parents' lessening tension.

"We've got five-minute commutes on foot," I wrote Bailey. Strolling up the campus's brick sidewalks to my office those balmy late-summer mornings, I pondered the path we might have chosen—a truly simple life. Stimulating day jobs close to home; at night, concerts and books; on weekends a small yard to mow, beer and cookouts, kids' sleepovers.

We'd become angry with Bruce for his slow pace, and he'd gotten annoyed with our putting the brakes on spending that would have made his job easier. "We can't afford the painters," I told him one afternoon. They'd begun priming the new sheetrock. He stared at me, his blue eyes like marbles. Bruce was around thirty, and this was his first big project on his own. He'd been the foreman for a major builder, a man we'd talked to about building us a home at Mossy Dell.

"You want me to fire them?" he asked.

"No—well . . . Bruce, we've run through our budget."

"You can't get good painters any cheaper."

"I know, but we can't afford them. We'll do it. Look, talk to us before hiring anyone else."

We hadn't had enough money, as it turned out, to move from Indiana to Ohio, purchase a house in town, buy two farms, tear down the cabin and Fred's house, build a new house on an old foundation, and endure the inevitable cost overruns of construction.

After work and on weekends, Kathy and I drove out from Bella's and painted the walls and ceilings. We painted slowly, steadily, silently into the night, our rollers a matching rhythm from separate rooms.

We moved to the hilltop in mid-October. One bathroom worked, a sheet of chipboard covered the stairwell to the second story, and the house felt rough without baseboards or molding around windows and doors; our oak trim was stacked in the tin open-front shed just past the outbuilding Fred had used as a heavy equipment garage. With the front porch unfinished, we entered the house on a springy gangplank supported by concrete blocks in the mud. Claire and Tom retreated to their bedrooms—big, freshly painted, and sunny—that we'd configured from the three tiny rooms on the south side of the house. Our bedroom was the former parlor off the kitchen.

We tried to shield my mother from our debacle, but it was impossible for us to be cheery when we talked about our new house. She sensed we were in trouble. I finally told her a few details. "Like father, like son," she said when Kathy took the telephone. It was true I'd rushed ahead before our residential needs had been resolved. For an instant, I remember, I actually felt proud that she'd compared me to my father. Still, even though Kathy shared blame and didn't accuse, this was *my* disaster.

Cardboard boxes were still packed from our move to Athens. Our winter clothes lay on hangers piled in the barn atop my open sixteen-foot trailer. The barn's doors had blown off years before, leaving our possessions at the mercy of storms. Mice ate holes in the linings of our coats.

"We can eat out at a restaurant once a week," Kathy said, announcing a strict budget. Claire and Tom looked glum. "The kids get to choose," she added.

To be so broke took some getting used to. I had to scrimp to save $20, and then I haggled with a rural grocer, who sold a few trees on consignment, to afford two spindly honey-locust saplings to plant west of the house for future shade. Drafts blew through our halls from the unfinished upstairs. As the nights grew frigid, Kathy declared we'd keep the thermostat on 60 degrees all winter to save money, and we stayed chilled. We finally hooked up Fred's old iron woodstove, disconnected during reconstruction, and could warm ourselves in front of its fire near the kitchen table. The stove wasn't airtight, and it consumed wood with fearsome greed.

I bought a used chain saw and on weekends cut firewood down the road in Mossy Dell's farmyard, where there was a chest-high stack of barky gray slabwood from the fallen oaks, and hauled the fuel up our hill in my truck. We bundled up, and slept under piles of blankets. Claire and Tom sensed our vulnerability and didn't often complain. I'd drive them to their schools each morning, feeling grateful for that one bit of normalcy, one routine that hadn't changed.

Unpacking my farming library one night, I shelved Louis Bromfield's memoirs beside Dad's old agriculture texts from Cornell and his own book. For the first time in years, I opened *Success Without Soil*, and noticed it was dated August 10, 1948: the twenty-eighth birthday of my mother. Dad had hired Rozelle Rounsaville as his secretary the September before. She couldn't have been more different from anyone he'd known. A five-foot-two redhead, and a Democrat, she'd grown up in a modest household of ten children in the sweltering country town of Atoka, in southeastern Oklahoma. Her father was a math teacher who eventually served as superintendent of schools and later, like Kathy's father, became the town's postmaster.

I'd always loved Mom's stories of my parents' California years—Dad used to fly her from San Diego to Palm Springs for breakfast in his own airplane—but now I saw Dad's expenditure of money and energy in a darker light. And I finally understood Mom's comment made years ago in our Indiana dream house: "You'll never put your heart into another place like you have this one."

In 1952, with Mom pregnant with Meg, Dad was running cattle on 6,000 acres of range near the San Jacinto Mountains, in Riverside County southeast of Los Angeles. Hydroponics hadn't caught on, and he'd given the business to his partners; he bankrolled the ranch with an inheritance of $300,000 from his father's estate. He named the property Atoka Ranch, in honor of Mom's hometown, and had contractors build a house she'd designed, as well as a barn and corrals. He bought registered polled Herefords, short-legged stock more suited to a show ring than to high-desert scrub.

Mom was appalled when he spent $20,000 for a bulldozer to clear sagebrush himself. She argued it would be cheaper to hire an operator at $100 a day. When he bought 3,000 acres of neighboring mountain land that was too rough for cattle, she asked him, "Why don't you get a herd of goats up there and let them clear it?" He seemed eager to burn through his father's money as fast as possible.

Early on, he saw that he'd picked the wrong place to be a grass farmer. With precipitation under twelve inches a year, it took more than a hundred acres to support one cow. He was buying water from the Colorado River to irrigate a mere twenty-eight acres for intensive forage growth.

Frequent guests on the ranch included his children with his first wife. One summer night Dad took Chuck and Ann on a hunt to kill jackrabbits, which were attracted en masse to his irrigated oasis. His ranch helper drove the pickup truck, and Dad shot a rifle from the passenger seat at jackrabbits frozen in the headlights. He threw the dead animals in the bed of the truck with his cringing children. Ann, then ten years old, was horrified by the slaughtered hares, and especially by the shrieks of the mortally wounded ones. Hearing her story fifty years later, I couldn't help but wonder if Dad was just trying once again to get in the rabbit hunt his father had promised him.

Mom divorced Dad three times during their early married years, over his black moods and her desire for children; after his divorce, he didn't want more kids. With each split, Mom gained his assent for another child. As she'd done early in her first pregnancy, before Dad knew she was expecting, Mom divorced him after she became pregnant with me. They remarried in Reno just before my birth, three months premature, in March 1955. "Pregnancy was toxic to my system," she always said, and remembered being violently ill in the truck. I think they were in Nevada buying cattle.

They sold out after ranching there for under four years—an eternity, I now knew, when bleeding money.

I'd vowed to Claire and to Dream's owners that once we moved to the hill-top, we'd fetch him. So one Friday I called and asked them to have him and his tack ready, that a friend was hauling him home for me on Sunday. The truth was I had asked Jim one day at the barbershop to take him to an auction. I didn't trust the horse and wanted him nowhere near our farm.

I met Jim at Maya and Bill's barn, where Dream waited in a stall. The morning was cool and damp, and Jim wore a short denim jacket that matched his usual Levis. He snapped Maya's yellow rope to Dream's halter and led him to a little dome-roofed red horse trailer. Dream balked at the door, and Jim, unsurprised, moved behind him and began to tap gently but steadily atop his hindquarters with a stiff white whip. It was the first demonstration of Jim's animal-handling skill I'd seen, and I was impressed by the way he slouched behind the horse as if he had forever, as nonchalant as if he were cutting my hair. Dream looked unusually awake, however; his head was up and so were his ears. There was no struggle or shouts, just Jim flicking at Dream's haunches, his spare hand tending the cigarette that smoldered at his lips. I stood back and watched this stalemate, recalling how Jim had told me that all animals, including people, seek to escape from pressure.

Bill came rushing toward us from his house. I'd never seen tall and gawky Bill move that fast, like he was late for a party. He stopped close to Jim, who glanced at him and nodded as he turned back to Dream. "You're the horse whisperer?" Bill asked. I was ashamed I'd told him about Jim using that phrase. Jim shrugged and kept tapping Dream on his wide brown rump. Suddenly Dream heaved himself aboard the trailer.

I'd told Jim to keep whatever he got for selling the horse as his fee. Jim said later that he towed Dream to an auction in Amish country and got about $500. I'd paid $1,000 for the horse and his tack, but the loss didn't trouble me: Jim had made one of my problems go away. I placed an ad in the *Athens Messenger* and sold Dream's saddle and bridle myself for $200—less than I'd asked, but something.

Two Fridays later, the kids and I traveled to a farm near Columbus to look at a dog being advertised for sale in *Farm & Dairy*, a Great Pyrenees that might provide an extra layer of protection for our flock. Unlike the iconic black-and-white border collies that chase and herd, guarding dogs follow the flock, living among the sheep and warding off coyotes with their territorial barks. Along with New Zealand electric fencing, such dogs were growing popular in America because they could replace the endless killing of predators with a kind of Cold War standoff.

On the hilltop across from Mossy Dell, we were beyond earshot of trouble. And more than coyotes worried me. I'd seen a tawny dog that looked like a pit bull cross roaming the neighborhood with a black curled-tail mutt. Their home appeared to be a trailer, halfway to Ernie's place on Ridge Road, owned by a man named Johnny who sometimes worked with Fred Paine; the rear of Johnny's property bordered a section of Mossy Dell's north pasture.

I paid $200 for the dog, whose name was Flower, reinvesting Dream's tack; the dog's breeder threw in an orange kitten for Claire in the bargain. Flower, who must've weighed over one hundred pounds, went limp when I tried to coax her into my truck's bed, so the guy helped me lift her inside under the topper. Claire cuddled her new cat all the way back to our hilltop. "Her name," she announced, "is Sunshine." In spring, skinny little Sunshine would deliver six kittens, the first births on our new farm.

In early November, six months before lambing, I bought a bare minimum of cheap hay, 150 square bales at $1.50 each, harsh with coarse stems, to get our small flock through the winter. I hoped the sheep could subsist mostly on stockpiled grass, ungrazed and unmown grass left standing, a compensation for having understocked pastures.

In mid-November, Glen Fletcher brought me Mister George, driving several hours from Dayton with the ram standing in the bed of Glen's shiny red pickup, confined by white-painted wooden racks. When we unloaded George in Massey's shed, the ram stamped a front hoof and tossed his head to show his concern. A few days later, Mike Guthrie helped me deworm him.

On Thanksgiving I turned George in with the ewes, well matured now at ten months. The girls ran from the dignified old male in exaggerated alarm; George ambled after them. I was concerned about my flighty females until I saw him breed a ewe two days later. As should've been apparent from Red's flirtations with me, sheep impregnation involves a real, nuzzling courtship, albeit brief. The ewes solicit a ram when they're ready. All appeared to be on track for April lambs from our fifteen ewes.

The flock grazed the pastures around Massey's shed until mid-January when a snowstorm hit, covering the grass, followed by freezing rain, more snow, and more rain. For the first time, I saw the sheep appear miserable outside in bad weather. They stood, heads lowered, on packed snow and ice as a thunderstorm dumped water on them. It was truly awful weather. They couldn't paw through the snow to get to grass, because a slick sheet of ice under the snow sealed off the forage. I threw them bales of hay. After seven days of this weather stalemate, at the cost of ten bales of hay—fifteen precious dollars—the weather warmed. The thaw permitted me to move my electric fence, which had been encased in ice.

I spent four hours that Sunday getting the flock across the farmyard to fresh pasture on Mossy Dell's south side. The sheep put down their heads and ate like they were ravenous after the dusty hay. I was tired and watched them for a while, taking satisfaction in their pleasure, and then descended into the farmyard. As always, the still farmstead, crisscrossed by shadows from overhanging branches, was hauntingly beautiful to me.

I was surprised to see Kathy pull in and park our van. I leaned into her open window as she sat at the wheel. She looked tired and pale.

"I've had it," she said.

"What's wrong?"

"We've got to leave," she said. "I'll start looking for a new job."

"*Now?*"

"We need to get out of here within two years, before Claire starts high school."

Her breaking point had been the schools' dismissal of students as the snowstorm had approached. Parents hadn't been notified—at least we hadn't been—and children had made their way to homes and offices as best as they could. Kathy had returned from a meeting to find Claire and Tom sitting at her conference table. We'd gradually figured out how to determine when schools would be delayed due to weather (one radio station gave last-minute notice); the unannounced early dismissal of our children was more than Kathy's feelings of vulnerability could bear. Yet her notion of fleeing was unlike her, my model of resolve. I felt a fresh stab of guilt over her despair, even as it confirmed my own fears.

"I just keep thinking, what if I hadn't been at work? What if we had been out of town? Our kids were turned out into a snowstorm!"

"God, we can't go through another relocation," I said. "Let's remember our original plan."

She looked blankly at me.

"We should sell everything across the road," I said, "and build here, at Mossy Dell. That's what we decided. Remember?"

She didn't respond, and even as I said it, the idea of building a house sounded worse than Kathy's notion of retreat, given what we'd endured. Neither of us wanted to have anything to do with house construction ever again. We'd barely survived financially, physically, emotionally. We were drained and unsure; we needed time to recover. We needed to stop making plans, to stop living in the future.

Kathy's uncharacteristic outburst was short-lived. The winter is a slow time, potentially a peaceful season, and we ceased talking about our uncertainty. We had the kids, our jobs, and a gritty house. And one day that snowy January, I held in my hands a slick new paperback edition of *RFD*, the press's editors having responded to my interest and returned it to print. I also had my dream, of course: a farm, about sixty-five acres on both sides of the road. And a pregnant flock of sheep that grazed the wintry hills.

In mid-March I seeded a legume into the barren cornfield. Beginning at daybreak, when the soil was still honeycombed by frost, I walked for hours up and down the field and turned the crank on the red plastic seeder strapped to my chest. After the first hour, struggling across the mud, the seeding felt a lot like work.

I thought I'd picked the correct species for the field's low fertility and sour soil: lespedeza. Known as "poor man's alfalfa"—how appropriate—it would capture nitrogen from the air and store it in the soil for grasses. More upright than white clover and with pointed leaves and pale purple flowers, it had been a pasture mainstay in the region before technology and greater affluence permitted the use of limestone and manufactured nitrogen.

A few days after my seeding, the sheep passed a milestone. I returned the flock to the spot in Mossy Dell's north pasture where they'd begun grazing stockpiled grass in mid-January after the storm. In my farm log, I recorded that the flock had gotten two months of grazing from the five acres. This was a modest victory, given the small size of our flock, but it was an accomplishment nonetheless to feed fifteen sheep in the depths of winter from pasture alone. Most farmers don't try such things. I had precious little experience, but I'd listened to Mike Guthrie and studied books and the *Grassfarmer*, all of which assured me that winter-stockpiled grass wasn't only cheaper than hay, it was better feed.

I anticipated our first lambing and watched the yearlings grow round as they grazed their way toward spring. One day as I petted Flower—she demanded a few minutes of affection daily before slinking away—I was watching the sheep on a slope above me and saw an animal urinate without squatting to pee like the other ewes. The urine came from what I'd thought was its outie belly button: the sheath for the animal's penis. Jo had sold me a wether, a neutered male, for one of my ewe lambs. I e-mailed her, hoping she would correct this embarrassing error, maybe give a partial refund. Her cheery reply was, "There is always one that gets by!" Buyer beware, indeed.

When I called Glen, making arrangements to visit him during his late-winter lambing to learn how he tagged and castrated lambs, he was amused that I'd been so ignorant as to buy a wether for a ewe. It *was* funny, which is why I'd told him. But my appreciation had become a show. Weary of being harvested for my naiveté, and sobered by my heedless actions, I saw how my family endured the consequences and lived with the risks of what I'd done.

Kathy spent hours at the kitchen table every weekend, consolidating our credit card debt and shifting balances from card to card, taking advantage of no-interest introductory offers. She played the companies like a professional, keeping us just ahead of interest payments by moving our accounts every month. In this way, although we owed $30,000, we never paid a penny of fees that would have crushed us. Kathy was busy saving us.

Yet monetary trauma makes people doubt themselves. In my guilt, I was afraid to comment on her heroic work, and she seemed to take no pleasure in it. I knew Kathy felt responsible for uprooting us, and I knew I'd dragged us into disaster. We could have settled so easily into our house in town, or bought a tidy farmette where we could have gardened, raised a few chickens, added a handful of sheep, and maybe even grown into farming. If only I'd stopped with Mossy Dell instead of forcing the purchase of Fred's house. All for that damn cornfield.

I grieved anew for the decade in Indiana when we'd been committed to that one place. My emotional reaction to starting over had shown me that I wasn't cut out to be a pioneer. I'd ignored my fears and my doubts about leaving all that we'd built, and wished I'd argued with Kathy about her desire to move. Then my own actions had clashed with my need for order, and with Kathy's practicality and financial conservatism. I wondered if self-knowledge must be acquired so painfully. By allowing my hunger for roots and my fantasy of farming to run away with us, we'd almost wrecked ourselves in a spectacular heap on these Appalachian hills. Maybe we *had* wrecked ourselves, hadn't narrowly escaped, and just didn't know it yet.

But I told Kathy that we'd also been incredibly lucky. Although we were upset with our young builder for cost overruns—and we felt he'd coasted at the end, thereby gouging us—Bruce had torn down a house and built a new one on its foundation between July and October. Experienced builders wouldn't have done it, especially with such limited labor, and many would have failed. Beyond the usual bruises and scrapes, no one had gotten injured in the deconstruction and reconstruction. And Bella had loaned us that house. Yes, we were in a hole financially, but we had good jobs and could recover.

As with Fred's field, the healing would take time. But it was almost spring; we'd have baby lambs then. And one day soon we'd be eating our own meat—starting with that peeing wether. An animal that seemed, by then, a gratuitous blow to my battered self-esteem.

PART THREE

WATER AND EARTH

In order to enjoy agriculture, you do not want too much of it, and you want to be poor enough to have a little inducement to work moderately yourself. Hoe while it is spring, and enjoy the best anticipations.
    —Charles Dudley Warner, *My Summer in a Garden*

CHAPTER TEN

This is a warm dry April you are missing, if it holds.
—e-mail to Mike Guthrie

IN EARLY APRIL, A LITTLE OVER A WEEK UNTIL LAMBING, WHEN I moved the flock to fresh grass one afternoon I noticed a small ewe with scours, the farmer's term for loose bowels. Like all stockmen, I was becoming a butt and poop inspector, forever worrying about the consistency of the manure issuing from my animals. These scours were black, not greenish like the sort of digestive upset I could imagine from spring grass. The ewe didn't graze, just lowered her head and let the grass brush her muzzle. Going off feed is another bad sign. An animal that refuses fresh pasture is sick.

I found her dead the next day when I checked on the flock during my lunch break. Flies were working on her dulled eyes. Death is part of farming, and I'd be professional and deal with it. Yet this was our first and felt like a bad omen. That evening I dragged her up the ramp of my trailer and towed it to the hilltop. I got a shovel and buried her in the charred spot on the lawn across from the house where our builder had burned construction debris.

"Did you cut her open to see what she was carrying?" Mom asked when I reported the death.

God no. I couldn't be *that* professional. The thought of seeing her cold dead fetuses sickened me. I felt guilty, somehow responsible.

Within days the other ewes seemed distressed, though with different symptoms. They mobbed against a fallen tree, lethargic, reluctant to follow me to new grass. It had turned muggy and flies seemed to be harassing them. When they did move, several lay down immediately with their necks stretched on the ground. They breathed rapidly with mouths closed in a shallow pant.

I sent a panicky e-mail to Mike Guthrie, who was in Guatemala, working as a construction supervisor for a Christian group building houses for the poor. "Welcome to the sheep business!" he replied. "Who knows why they die? There is always a reason but it's not always easy to know why and seldom worth the cost of figuring it out." He ticked off a daunting list of diagnoses for the dead ewe. A fetus might have died and poisoned her, he said; most likely it was ketosis, the toxemia that all mammals, including human, are heir to: a female breaks down her own bodily reserves to feed a rapidly growing fetus. The ewe's flockmates could be suffering from bloat, Mike added, or a case of grass tetany, an obscure ailment arising from the sheep's calcium-magnesium balance.

Mercifully the flock shed its mysterious distress. They could have been touched by the edge of major illness or just feeling hot and fly-bit. I'd never know what killed my ewe, and never know what ailed her moody sisters. Maybe by then I was just watching them too closely.

The thirteen remaining ewes and the fat wether were grazing around Massey's shed. At the start of lambing, I would pull up all the electric fences and let them wander where they would. This was so the ewes could pick a spot in which to give birth, their lambing beds. There they could stay for several days with plenty of feed nearby and bond with their lambs. At least that's what the experts on Sheep-L said should happen. The five acres around the shed were enough to lamb twenty-five ewes, so the flock was in good position for at least a month.

One day I noticed a protrusion from Runt's vagina, maybe a dreaded prolapse. This can occur, I'd read, when the strain of pregnancy meets a genetic weakness and flesh gives way. Her tissue retracted the next day, temporarily. Having grown and lost her tarry butt, she retained the spindly appearance that had always troubled me. Her head, with a prominent Roman nose, looked too big for her bony body. On the flock's first possible due date, April 20, I recorded in my farm diary that Runt's prolapse was "huge."

I called a vet supply company and ordered a harness to force the flesh back into place; farmers did their own vet work—Mike had made that clear. But Kathy and I couldn't catch her and only exhausted ourselves running around the pasture. She could really move, especially uphill with all four hooves helping. Bipeds hadn't a chance of grabbing her. She lured us into sweaty trembles by sitting down after she'd accelerated. We'd creep back, lunge, and the race would begin anew. The blue nylon halter in my hand was a joke. I wasn't sure what to do anyway except get her in the shed, try to push the tissue back into place, and pray it held.

The first yearling to lamb did so exactly on the first due date, calculated from when I'd turned in George. That afternoon I found she'd gone to the far end of the pasture and dropped twins, both females to expand the flock—good ewe! She didn't bring them to the flock for two days—great ewe! This is what I'd read a great maternal ewe should do. I jotted down her mothering score, four out of a possible four points, and her physical description, a medium-sized white ewe with a black spot on her left ear. Her ear tag identified her as 811, but she'd earned the sobriquet The Good Mother.

Tom and I arrived the next day almost in time to see the second ewe deliver her lamb. She acted like she'd just dropped dung instead of a baby. She walked away and grazed in the warm afternoon sun, leaving the poor lamb flopping, covered in glutinous birth tissue, in the thick spring grass. Looking at The Good Mother, standing like a statue with her pair of lambs, I wondered if this was the variation I could expect. Since buying from Jo, I'd corresponded with shepherds on Sheep-L, and had concluded I'd been a fool to buy from someone who lambed in a barn, instead of getting ewes from a flock that had been selected for mothering ability on pasture. The clueless ewe, now known as The Bad Mother, might have functioned in a small pen in a dry barn, but she was disastrous outside, where maternal behavior is more vital. Eagerly I condemned the ewe, my inexperience, and Jo's incompetence as a breeder. This was another emergency.

The Bad Mother hadn't moved away from the flock to give birth, so the other ninny yearlings took turns running to see her lamb, which tried to follow and nurse them. Our new guardian dog ran around, distressed; she knew where the lamb belonged and chased the gawkers away. Flower's concern and her attempt to resolve the situation impressed and amused me. But the lamb tried to bond with *her*. The hungry lamb even tried to nurse her, which confused Flower, so she lay down. The lamb snuggled against the fluffy white dog. The Bad Mother, grazing nearby, would baa to her baby once in a while and then butt away the poor creature when it ran to her and tried to suckle. I felt helpless, unable to answer Tom's question: "Is the mother going to let Flower raise her baby?" I drove Tom to the hilltop and asked Kathy to help me before we lost our light.

We nudged The Bad Mother into Massey's shed and I grabbed her neck; Kathy got behind her and hugged her midsection. Without a stanchion to hold her, I improvised, tugging my blue halter around her ears; with a lead rope I pulled her head through an opening in a welded-wire fence panel. I tied her so tightly she couldn't move. Hard for her, but the lamb could nurse and she couldn't stop it. Night was falling.

"Won't she hang herself?" Kathy asked.

"I hope so."

"We can't leave her like this."

Kathy was right, in that the ewe might crush her offspring if she collapsed. I didn't want to untie her, though, and risk her rejecting or even hurting her lamb. So I jammed a bale of hay under her front legs, leaving her udder exposed. I hoped the dense cube would keep her upright if she fell or tried to lie down while tied. Her lamb baaed and toddled over, nursed, and lay against the hay.

As I drove us toward the hilltop, someone crossed Snowden Road and stood waiting on the shoulder on my side of the truck. It was Sam Norton, who lived in the tidy gray clapboard house at the corner of Snowden and Ridge roads across from our hilltop. I'd met him the previous year during my hectic comings and goings from town, and had given him one of the daylily plants we were transplanting to Mossy Dell. He always waved when I passed his homestead, one of the notorious two-and-a-half acre lots Fred had sold off. From our house we'd see him mowing, raking, pruning, washing cars, painting, and puttering endlessly around his house and two white outbuildings. He was retired from working for thirty-four years as a deliveryman at the university, where his wife had served food in a cafeteria.

I stopped the truck and rolled down my window, wanting only the peace of dinner and a beer—maybe two. Sam grinned at us, his hands fretting at the seams of his trousers. At his feet rested a dirt wedge topped by a bunch of daffodils, still sporting bright yellow blooms.

"I dug you some Easter flowers," he said, picking up the dirt clod with its cap of greenery. "Erma and me thought you might like some for your place."

"Thank you," Kathy said. "They're pretty."

I got out, lowered the tailgate, and Sam set the clump inside.

"How many lambs you got so far?" he asked me.

"Oh, a couple. Lambing's just started really."

"I saw you go by the other day with your trailer. Is everything okay? Need any help?"

So he'd seen the dead ewe. "No, I'm fine."

As I drove away I said, "He's going to be a pest."

"Richard, he's trying to be neighborly."

"He's nosy."

"You're touchy."

Kathy went inside to make dinner, and I got my shovel and planted Sam's daffodils on the edge of the bare spot where our first casualty of lambing lay buried.

In the morning I approached the shed, the sky just pinkening over Lake Snowden. I dreaded finding death. Instead, The Bad Mother had had a second lamb. It appeared that a lot of thrashing around had occurred— the bedding was churned and the hay bale stuck out from under her body at an angle—but my tether had held her close. Neither lamb had been cleaned, their coats crusty and crisscrossed with strings of dried amber placental tissue. Yet their eyes were bright, their bellies rounded with milk.

I untied the ewe, and she sniffed at both babies and nudged them toward her udder.

I knew from Sheep-L debates that some shepherds would give her another chance; it was her first lambing, after all. Some would cull her after weaning and keep her ewe lamb, because the lamb was only half the mother's genetics. I found I couldn't focus anymore on such talk, though I'd always loved it. Problems were real, and too many were happening at once. Nothing was stable or orderly or clear.

On Sunday afternoon I got another ewe into the shed, one of the nicest yearlings, a big white sheep and roly-poly in pregnancy. I asked Kathy to look at her because I thought she'd been in labor a long time. "She's in trouble," Kathy said. "I think you should call a veterinarian." I didn't want to seek help: real shepherds avoided paid assistance. But I dreaded starting the work week with a vulnerable ewe. And emotionally I didn't think I could handle more death. I called the closest veterinarian, Becky Lanier, an elfin woman who parked her truck in the farmyard and strode uphill to the shed carrying a box of tools, followed by all of us. In the shed in the fading afternoon light, the ewe lay on her belly, her sides bulging like saddlebags, her neck outstretched. I knelt beside the vet as she lubricated her hand and forearm and inserted her arm past the wrist. "I feel a lamb," Becky said. Kathy ushered Claire and Tom close to watch the miracle of birth and they stumbled forward.

Becky pulled out the lamb; it was big and white. And it was dead, crumpled on the soft hay I'd spread over the petrified manure floor of Massey's shed. The vet inserted her arm again. The rest of us looked at the dead lamb in silence. "I don't feel anything else," she said. "Here, lube up and see if you can."

What? Putting my arm inside a ewe was the last thing I wanted. *This is what I hired you for.* I was a grass farmer, not an obstetrician. Yet I was embarrassed by my squeamishness over slime and blood and birth fluids. Becky was teaching me about an inescapable aspect of livestock farming and the vortex of mammalian life. I squeezed out lube. Kathy and the kids moved back. I felt my way into the squishy darkness, but hit nothing solid. I went deeper, wondering how far a person could stick his arm into a ewe.

"I don't feel anything," I said.

"I guess that's it," Becky said. "But you better keep her in here overnight."

The next day, I checked on the ewe during my lunch break and found her suckling a lamb. She looked pleased with herself, as if humans had been thwarted again. Although we hadn't found that lamb, Becky had saved it by removing its dead sibling.

I saw that Red had lambed—twins—and walked over to see her. "Red! Good girl," I said. She hadn't gone as far from the flock as had The Good Mother, but was staying close to her lambs and wasn't dragging them around. Red looked at me and quickly sniffed the butt of each lamb (*Mine. Safe.*) and nudged them toward her udder. They began to nurse, one on each side. The calm rightness of the scene contrasted with the tension of another ewe who'd dropped a single lamb that morning; she eyed me suspiciously, tensed to flee, and I backed off.

The next morning, Red let me tag her offspring. She stood beside me with mild curiosity and grunted to her lambs flopping in my lap as I pierced their ears with numbered tags. I sat cross-legged in the grass and tugged out each lamb's right ear, inserting a plastic tag with special pliers. Lambs were easy to catch within twelve hours of birth unless their crazy mother fled upon my arrival. When I'd tried to nab the wild ewe's lamb, she had run from me as if I were a bloodthirsty wolf, and her lamb followed, mimicking her mother's panic.

I checked on Runt and found her bloody at her rear and with vaginal tissue hanging to her hocks—I'd forgotten to ask the vet about her! Her troubles had dragged on for a week. With newborn lambs mixed throughout the flock, and the ewes scattered across five acres, trying to run everyone into the shed to catch Runt wasn't an option.

By the next day, Saturday, Runt was stumbling as she walked, though her prolapse looked better. On Sunday morning she was down, sitting with her legs folded beneath her, and I could smell rot from forty feet away. Otherwise she appeared normal and unconcerned. The flock showed no interest. And strangely neither did Flower, who otherwise anxiously monitored each ewe.

Unable to catch her to treat her, and certain that she'd suffer and die, I felt that I must kill her to spare her further suffering. I got out my .22 rifle that afternoon, sat down in the grass thirty feet away, and shot her through the neck as she looked at me. She gracefully rolled over. I put another bullet into her forehead from a few inches. I saw that she'd pushed a rotten lamb halfway out. It had been dead in her womb for at least three days, an unseen problem separate from the prolapse.

I went and dug a hole for her in the scorched earth across from our front door, my sheep graveyard. I tried to suppress my anxiety and sadness, which I couldn't afford and was reluctant to admit to Kathy. As I shoveled I thought how stupid I'd been to buy Runt, and that having done so, I should've butchered her in the fall and eaten her. Trying to be a farmer was so different from what I'd been: a suburbanite with a hobby farm who measured success in emotional terms and wrote naive gardening columns for the newspaper. *Look, I raised a ripe tomato. Aren't I cute?* I was taking the impersonal difficulty of farming very personally. Stabbing the tight dirt, I thought of our manicured Indiana acreage, our gracious custom house.

This was so *hard.*

And Runt had been weak. Nature isn't gentle as it tests strong and weak alike. Lambs, born in cold dew, butt ewes' udders violently as they nurse. The ewe grazing on a sun-drenched day is ingesting parasites that will riddle her gut and bleed her. She's leaving behind worm eggs in her droppings that will hatch into larvae that will kill her susceptible lambs. The gentle breeze rippling the grass is drying the soil for the looming summer drought.

On a mild day, the birds singing and the grass soft under a cloudless blue sky, Flower was agitated when I checked the flock. She rushed up and led me to the sound of a lamb crying. I thought it had fallen into a groundhog hole or was deep in a multiflora rose thicket: I could hear it but couldn't see it. The lamb was somewhere on the brushy hill behind Massey's shed.

Finally I found the baby, trapped under a fallen tree trunk; the lamb had slid under the log and couldn't stand to extricate itself. Until I was standing over it, I couldn't see it. Now I could do something positive as a shepherd, save a life. I knelt and grabbed the lamb, and saw its mother— The Bad Mother!—grazing obliviously below. Her lost lamb sprinted across the pasture and joined its sibling. Both nursed frantically. The Bad Mother regularly lost track of her offspring. She went baaing through the pasture several times daily, sniffing other ewes' lambs and disrupting everyone.

I got in my truck and decided to check the cornfield on my way out to see if any of the lespedeza I sowed in March was growing. I was just about to pull over when I saw Sam at the end of the road, across from his house, standing there like a troll waiting to levy his toll. I drove on and stopped beside him.

"How's Richard doing today?" he asked. "Is Richard enjoying this fine weather and keeping out of trouble?"

I'd begun dreading Sam's third-person greeting, which caused my jaw to clench. *Richard's just dandy*, I wanted to spit. *How's good ole Sammy?* This was always followed by a wave of guilt. There seemed a hostile levity in Sam's cloying address, yet I knew my irritation arose from my tension over the sheep. My compromise was to ignore his pushy cutesiness as if he hadn't spoken, and my replies sounded flat and brusque in my ears.

"What's up, Sam?"

"I made some cornmeal mush," he said, holding out a square of translucent waxed paper tucked around a yellow chunk and clasped with a red rubber band. "I sliced it when it cooled. It's good fried in the skillet. I eat mine with syrup and butter, or peppered and salted with my eggs."

"Thanks, Sam, I'll enjoy it," I said, taking the packet. "I've been meaning to bring you some eggs. This time of year the hens are laying more than we can eat."

I passed the cornfield several times a day but didn't walk into it. The field was a dull green tapestry of weeds.

The last afternoon in April, Brownie delivered two nice brown lambs. All our other lambs were white, a few with spots. Kathy, Claire, and Tom hurried over to see the chocolate lambs, a male and female.

The next morning I flipped Brownie's baby ram on his back to slip a rubber band onto his scrotum with a special tool. Glen Fletcher had shown me how to perform this bloodless castration, which turned rams into wethers so that they'd grow toward market unable to breed their mothers or other ewes. The band stopped circulation to the scrotum and testicles, which gradually dried and one day sloughed off. If banded within twelve hours of birth, a lamb didn't seem to feel the procedure, and flies weren't attracted by a bloody wound.

In the case of the brown lamb, however, Kathy intervened. "He's so pretty," she said as I sipped coffee back on the hilltop. "I think you should go take off that band. Someone might want him for breeding." So I returned, ran him down, and carefully sliced away the thick green rubber ring. I doubted Kathy's prediction, yet respected her certitude, impressed that she was thinking ahead to our first sales. I couldn't see more than a day in front of me.

Lambing season ended that weekend with two more uneventful births. The final tally included two dead ewes, a ewe that didn't lamb, and a stillborn lamb. And eighteen offspring from eleven ewes. There was also the wether; we'd get around to eating him, along with the barren ewe.

The earliest-born lambs began to caper and race each other and butt heads; they tormented the ewes, pouncing atop their sleeping elders. Mister George lay around looking utterly exhausted by what he'd wrought. Kathy and the kids visited more, and the ornery lambs made us all laugh when, hopping forward, they suddenly danced sideways, tossing their heads, overcome with feeling good. But the funniest thing lambs did (one that would never fail to make me laugh over the years, no matter how many times I saw it) was when they approached their resting mothers, stopped, and baaed loudly, with increasing desperation—*Come to me!*—like human kids sitting in front of the TV, calling for mom to bring them snacks.

I was watching the lambs play in the sunshine one afternoon in front of Massey's shed when Mister George walked past and stepped into a loop of twine I'd used to tie the shed's gate closed. The twine wrapped tightly around his hoof, and George tried to escape by attempting to pull off his leg. I pictured his severed hock hanging from the gate, grabbed my pocketknife, and lunged forward to cut the string.

Maybe Mike was right: sheep *are* just animals looking for a place to die.

No measurable rain had fallen from mid-April to mid-May. It was becoming unbelievably hot and dry for so early in the year. Even so, the grasses headed out, determined to make seed. By June I was mowing pastures every weekend, and workday evenings until eight o'clock, trying to set the grass back to a vegetative stage for the flock to graze. Clouds of pollen marked my path; it constricted my throat and I coughed. My eyes itched and the tissues around them swelled; my neck broke out in a rash. Bloodthirsty deer flies flew up from Lake Snowden to cling to my scalp and bite.

And then a hurricane of locusts emerged howling from the earth.

The insects—seventeen-year cicadas, actually—were as large as late-summer grasshoppers, with chunky black bodies. They had fiery red eyes, clear wings with prominent orange veins, and seemed to fly intentionally at me as I mowed. The diesel thrum attracted them, my John Deere an improbable green decoy. After almost two decades underground, the females sucked juice from woody stems, mated, and deposited their eggs in slots they trenched into the twigs. The males gave voice to the woods surrounding Mossy Dell, as if the trees themselves had erupted in a buzzing roar. The wall of sound chastened songbirds into an uneasy silence. Local newspapers printed articles about the alarming emergence of so many insects—hundreds of thousands to each acre. Entire neighborhoods in town were draped in cheesecloth shrouds to protect young trees and shrubs from the onslaught.

One evening in early June, as I ran the mower along the edge of the pond in the farmyard in a few tidying swipes, a female mallard flushed in front of the tractor. Her furious wing beats and kazoo quacks in my face startled me. She'd held until the tractor's front wheels rolled almost on top of her. I stopped the machine, climbed down, and found her nest on the sloping bank. She'd lined the warm little bowl with dry grass and tufts of gray down pulled from her breast. I counted seven pale green eggs.

There was no telling whether she would return or leave her eggs to rot. I surmised that she was a young duck in her first nesting season, because she'd picked a poor spot—beside a tiny pond without much food or cover—instead of along the nearby shoreline of Lake Snowden. She'd have to lead her brood to the lake, and would surely lose ducklings on the way. I knelt on the slope and thought. I'd always wanted to raise some real mallards, sleek and shining—so unlike hatchery birds with their faded colors and bodies thickened by domestication.

I pulled off my sweaty baseball cap and placed each egg inside with my fingertips. I would carry them to our house across the road and warm them in a little incubator I'd bought for hatching chickens. Maybe they would bring Mossy Dell's ferny magic to the hard ground of our hilltop.

In the middle of June, my youngest brother, Pete, came from Florida to help me fence. We got out the post driver I'd bought from Fred Paine, attached it to the tractor, and began to pound. We had no faith in the implement's ability to drive thick posts into powder-dry clay.

Once the pounding started, we were shocked by the sheer brutality of the process. It took two full days before we believed in the device, despite its having plunged scores of posts deep into the gnarly ground. What it did didn't seem possible. We drove blunt corner posts six inches in diameter three feet through tight subsoil. A small pounder wasn't much faster at getting posts into the ground than the traditional tractor-mounted auger, which digs rapidly. But when you were done, you were done: no hole to fill with dirt and tamp by hand. And the driven posts wouldn't budge for man or beast.

We finished a corral I'd started in front of Massey's shed as Mister George's bachelor quarters. Then we went to the hilltop and began a second corral behind the big red barn. We never got comfortable with the pounding. Not that it was particularly loud—a stunning *thwack* followed by a high ringing—yet it was harsh and unnerving.

Pete held the posts upright and in position with a short steel safety hook I'd ordered from the pounder's manufacturer. Then, sitting in the seat, I dropped a lever and the tractor's hydraulics forced a piston downward to smash the post with a force of 30,000 pounds. A wave from the impact rippled up Pete's arm and jerked his shoulder; it looked like my brother, a burly police homicide detective, was firing a magnum revolver. My tractor, which weighed 1,500 pounds, bucked as it recoiled from each blow. Bravely holding the posts with his little hook, Pete was in a more intimate position than I was in the tractor's seat. He admitted it was scary.

"In the blink of an eye," he said, wiping sweat from his brow with the back of his hand, "I could become Flat Hand Luke."

In the midst of our project, I left Pete in bed early one morning and rushed over to Mossy Dell to meet an excavator. When we were still living in town I'd called him about renovating the farmyard pond, approved by the Farm Service Agency for federal assistance. The pond appeared fine until you stood on the dam and looked down. Water overflowing in heavy rains had sliced into the dam's backside and cut a deep gully across the woodsy pasture below. My idea of repairing the problem seemed to have been made a lifetime ago, and I thought of backing out. But this guy was a legendary pond builder. He was the best excavator in Athens County, everyone said. And almost impossible to get, they added.

When I pulled into the farmyard he was already there, sitting in his mud-spattered white Ford truck with bars welded over every running light. Daniel was his name, and in the formal, olden-times way of some country folk, he didn't abbreviate. He turned out to be a slight, desiccated-looking old man—he told me he was seventy-four—wearing a white shirt with mother-of-pearl buttons. He seemed cheerful, but I sensed in his wiry body and ice-blue eyes a seam of Scotch grit.

I'd heard his bread-and-butter work was helping build bridges along the Ohio River and constructing retaining walls anywhere a hillside gave way and dirt covered a road; then the highway department called him, and landowners had to wait in line. But another thing made Daniel attractive: he sometimes didn't bill landowners, or so it was rumored. Maybe he was so busy that his billing department (probably his wife) couldn't keep up. Or, since farmers couldn't truly afford him, his crew, and his equipment, he built their ponds as a public service.

There was another story, a sad one. Annie Clark, the dairywoman who'd sold me Jack, told me that Daniel had killed a girl. Decades ago, she said, Daniel had gotten drunk one night and crashed his vehicle into hers. He'd probably put in a fifteen-hour day and then had a few drinks with the crew; if it was just beer, a case on his tailgate, they wouldn't even have considered that drinking. But Appalachia's curvy roads are unforgiving, even if you're sober. The judge let Daniel off easy, Annie told me—no jail time. Even then, as a young man, he was admired.

I was startled to remember this tale about him as we shook hands. In the coolness of the morning the locusts weren't yet roaring.

"There were a couple dogs in your farmyard," Daniel said.

"Two?"

"A big white dog and a black dog with a curled tail. They were playing."

"The white dog is mine. She's supposed to run off strays."

"I'm glad I didn't shoot them. I thought about it. I could hear your sheep up the hill."

We inspected the pond and descended toward the gully below. "I know this land," Daniel said. "When I was a boy, I dragged raccoons in a gunny sack behind my pony all over these hills, leaving scent for hounds."

"So you knew the Vaughts?"

"Yes. They hosted the dog trials, had a cabin here." He looked up to the empty space where the cabin had stood. "You live up in Fred Paine's old place."

"Yeah, I'm split between two farms. This summer I'm fencing here and over there. We're driving posts with Fred's old pounder, and it's slow."

"Take a chain saw and sharpen the ends of those posts. Try to make the point like a pencil, even all around, or they'll go crooked. And be careful. My foreman got two fingers torn off last fall by my pounder. We were ramming pilings down on the river and I dropped it too quick. I thought he was ready. The doctors sewed his fingers back on."

We walked the gully. I didn't mention the livestock watering system we'd discussed, and then he brought it up. Daniel ran cattle himself, so he knew I needed water; I'd been hauling it from the hilltop in my truck. We went to see the pastures, climbing uphill on the overgrown roadbed that ended at Lake Snowden, and came to a metal gate, overgrown with multiflora and rusted shut, five feet high. We paused—I thought—at a natural place to catch our breath. I was gazing across the farm when from the corner of my eye I saw Daniel simply melt over the gate, quick but unhurried, fluid and silent.

He'd shown me a rural skill I hadn't even known existed. He must've defeated many such hurdles during his days and nights roving these hills. And yet there was something beyond skill involved: it was as if he'd entered another dimension before my eyes. I wanted to see it again. I knew how *I* climbed the farm's arthritic gates: slowly, precariously, and with flailing, middle-aged effort.

And incompetently, I now saw.

Daniel was older than me by thirty years. I mounted the barrier after him with my earthbound clumsiness, which now felt like a deeper flaw.

One morning in late June, two eggs were chipped in the incubator. The kids and I could hear murmuring peeps, see the tips of dark bills working to enlarge the holes where the eggs were pipped. I called to the ducklings, which answered with shrill urgency from inside the shells. We'd have wild ducks. I felt blessed, and felt I owed these creatures. They were conceived in living waters, brooded to life in the hemlock-shaded coolness of Mossy Dell's old farmyard, and carried to a Styrofoam box inside a construction site on a sun-blasted hill.

One thing was clear as the biddies emerged from their brittle cocoons, wriggling like wet seals from the rocks: these were some sweet ducklings— literally; they smelled like maple syrup. I'd misted the eggs daily with water during incubation, using an old pancake syrup bottle as a makeshift sprayer, and the incubator's warmth had reconstituted a residue. The sugary scent passed through the eggshells and coated the ducklings. All seven eggs hatched, and when the brown-and-yellow brood huddled in our children's laps, the room filled with the smell of Sunday-morning flapjacks.

There was another odd thing about those wild birds. They were sweet-natured, the tamest ducklings I'd ever known, and I'd raised a lot of ducks in my boyhood bedroom in Satellite Beach. Unlike baby chickens, ducklings are afraid of their caretakers. Their terror had distressed me as a boy. Domestic ducklings, gently tended, fled my hands—no matter how slowly I moved as I gave them feed and water—and threw themselves against the walls of their brooder. Yet these mallards talked to us, and nibbled our fingers with their cartoonishly oversized bills.

Then I remembered that during incubation, uncertain of when the ducks would hatch, I'd talked to the eggs in a high-pitched sing-song as I misted. "Baby ducks, baby ducks. Come on, baby ducks. When will you hatch?" Inside the eggs, the ducklings had listened. They'd imprinted on my voice.

They made terrible messes in their brooder, a cardboard box in the kitchen, gleefully soaking their newspaper bedding and dissolving their dry mash into mush in their waterer. With their downy silliness, they softened the scorching summer and the rawness of our makeshift home. I was too busy to play with them, but Claire and Tom weren't. They held and petted the lucky seven, took them swimming in our bathtub. Claire's horseback-riding friend Lucy liked to lie on our couch and watch television with the fluff balls nestled on her abdomen.

The mallards, a surprising success, grew up—rapidly, as ducks do—and began to spend their days in the little pond in the hilltop's south pasture. They stayed remarkably tame and ate at the barn with the chickens and guineas. I cherish a photograph I took when they were babies, slumbering in a row on a blurry white field: Lucy's belly.

I took another photo that year, one that tells a different story, one that reminds me of all that we'd come through—Kathy, the kids, and I—to arrive at that first lambing. Claire and Tom sit behind Massey's shed at the base of the hill, which rises behind them and meets a line of trees and brush along a boundary fence. The trees, stiff-branched, haven't yet leafed, but the shrubs bristle with new growth. The grass appears soft and full, as billowy as a blanket, beneath the children and across the ground. That spring was warm and dry, and of course the summer would be scorching, yet in that moment at the end of April the pasture retains the emerald lushness that for a time every year makes southern Ohio look like Ireland.

Claire, days before her twelfth birthday, her brown hair on her shoulders, wears a long blue dress sprinkled with a pattern of white flowers. She holds a white lamb in her lap. She and the lamb are both looking at the camera, Claire with a big grin and the lamb intently, its dark eyes as eager as a puppy's. I'd grabbed it from the pasture for the photo, and it's looking for its mother.

Tom, nine, sits cross-legged beside his big sister. He still hasn't learned to smile at the camera, but tries, his mouth pressed into a downward line that bunches his pink cheeks. Tom wears a blue T-shirt with white bands, and he must have been in a growth spurt because his canvas pants are too short and ride up his legs. He scratches at his neck with his left hand, bothered by his long hair, which forms a dark blond helmet on his head and hangs down his neck and in his eyes. His little face peers out as if from under a haystack. In Indiana I'd kept him barbered regularly; here, our Saturday barbershop ritual often collapsed, a casualty of house construction and farm busyness and living in the country and the unpredictable weekend hours of our Appalachian barbers.

The memory of that first lambing still lingers, like a vivid dream in my mind's eye. A chaotic scene: white ewes drop lambs across a green pasture, and some butt at lambs—rejecting their babies! Far from having beginner's luck, never again would I experience as many problems at once.

When I waved the kids into place that day for their portrait with a lamb, I wanted to capture a culmination, and I suppose I did. Now I can't look at the photograph in its cherry frame on my desk without seeing something else. I knew Tom needed a haircut, but I didn't notice how much his hair was bothering him, or sense how his unease epitomized our unsettled lives. That's the poignancy of family photographs. We try to memorialize something, and instead document what we can't quite see and don't yet know.

CHAPTER ELEVEN

I find myself fantasizing about a condo in Minnesota, the truly simple life.

—Farm Diary

In late June, Kathy left for Hong Kong, where she was teaching a class to help pay off our house debts. I'd assured her that pond work and water lines would be subsidized for at least half their costs by the federal government under conservation and grazing-incentive programs. In truth I was hazy about what I was getting us into financially, because Daniel cost about $1,000 a day, and he'd said the job might take three days. I hoped that, with the federal help, I'd have to pay him at most $1,500—a small fortune to us by then. I don't know why Kathy didn't protest; maybe it was her affection for good deals. Or her old guilt for having moved us.

With her gone, I took a two-week vacation from the press. I'd care for Claire and Tom and configure farm and animals for a farm sitter, and then drive the kids to Florida to see their grandmother. Daniel would work on the pond when I returned.

But he called right after Kathy's departure to say he'd be able to start the following Monday, the first week of July. When an excavator of his stature was available, you dropped everything. No problem, I thought. The pond will take two days, the water lines a third. I'll still have time to seed the bare ground and scatter straw on top. I could hold to my schedule, if barely, and be done in time for the trip with the kids. I'd worry later about the expense.

Monday went well. Daniel and his crew bulldozed the gully below the pond and buried a French drain the length of the field. A defunct 400-gallon water tank was in the way, and they jerked it up with their track hoe, a giant ice-cream scoop mounted on bulldozer tread, and repositioned the circular concrete vat as easily as you'd set down a poker chip. I worked on replumbing the tank for the sheep.

On Tuesday, the rain we'd been praying for arrived, a brief but heavy soaking. I worried about the earth we'd exposed washing away, but the dusty ground absorbed the downpour. With the site turned to mud, Daniel declared the day rained out.

On Wednesday, they finished grading the field below the pond, shaping a gentle swale above the French drain, and they repaired the emergency spillway on the back of the dam. Excavators always need more dirt, and Daniel filleted it with his bulldozer blade from the woodsy pasture above the drainage field. They knocked over trees, peeled away the sod, scraped off tons of dirt, and then pushed back the topsoil so that the pasture could again grow grass. They shoved the remains of trees into bristling piles. I wanted to close my eyes—such beautiful trees—and yet the destruction was also oddly thrilling.

A thunderstorm threatened all day. The sky to the west was black and it rumbled, an evil presence. I felt vulnerable thinking of what a big storm could do to the raw soil on the dam, and to an acre of bare ground below the pond and in the neighboring steeply sloped pasture. I saw the folly of average weather: a normal year is made of sudden changes and extremes, which in their wild swings produce averages. The weather lurches from drought to deluge, from balmy to frigid. And in the end, a grinning weatherman says we received average rainfall, even if we got flooded all winter, and during the summer, when we needed rain, vegetation burned to a crisp. This summer even people who didn't pay attention knew the heat and dryness were extreme. And the relief valve had to be a thunderstorm.

In late afternoon, Daniel's radio reported that high winds were hitting

Athens, twenty minutes north, with vicious force. Mercifully, we caught only the edge of it. The light rain cooled us and then we steamed in humidity that wouldn't lift. The pressure was still building. I was distracted from my unease by how dirty I felt. And how tired. I went home that evening, fed Claire and Tom a frozen dinner Kathy had left, and collapsed.

When Kathy called late that night from Hong Kong, I launched into my tale of work and heat and thunderstorm terror, about how complicated things had gotten just before my vacation with the kids.

"I think I can make it, if nothing goes wrong," I told her. "If Daniel finishes this week and I get everything put back together next week, we'll make it."

"And how are the kids?" she wondered.

"They're foraging."

On Thursday, we trenched water lines in an ungodly heat, the atmosphere hazy with humidity. I couldn't believe the conditions, which had gone beyond oppressive and into hellish. I looked wonderingly at the men—how bad did the weather have to get before they'd quit?—but there was nothing to do but keep working. The radio reported the heat and humidity index at 110 degrees. I broke into beads of sweat the instant I moved, and got soaked as I unrolled black plastic pipe. The locusts buzzed madly in the trees, making up for their seventeen years of mute burrowing. The forest itself seemed to emit an angry wave of mechanical noise.

A scrawny, shirtless young man with a mop of sandy hair rode a trencher, its blade churning three feet into the ground through baked soil, roots, and a layer of shale; it spit the pulverized mass to the side. The trench was bone dry. I squatted, greasy with sweat, in the trencher's floury wake and connected pipe. Sweat rolled off my face into the dust. Cicadas screamed. Wilted grass crunched underfoot. The trencher, shuddering and disgorging a gluey diesel odor, radiated its own heat. The kid stuck inside became woozy, then faint and disoriented. Daniel pronounced him a victim of heatstroke and shut off the machine.

His other man, a quiet black-haired fellow in early middle age, completed the digging with the track hoe's big scoop. He was the guy whose fingers had gotten ripped off (he was shy about giving me a close look at the reattached digits). He obviously loved Daniel, who'd attended to him, he said, and kept him on full pay through his long recuperation. His bucket work was fast but made a wide ditch—more erosion to worry about on hilly runs.

Late that afternoon, Daniel and his men pulled out, leaving me wondering how to bandage all the gouges they'd made. The hills had turned my legs to rubber. My feet throbbed, and I moved in slow motion. My shirt was so wet I could see through it; sweat saturated my leather belt and soaked my pants to the crotch. I went home to the hilltop, fed the kids, and returned.

Protecting the steep pasture was critical. We were overdue for a storm; the heat just couldn't continue unabated. Riding my tractor and fighting a feeling of desperation, I dragged my disk harrow over clods and then strapped on my red plastic seeder and cranked orchardgrass seed onto the ground. I tore apart bales of straw, which would absorb the force of rain and runoff, and spread the shiny yellow mulch past dark, my nose clogged from dusty chaff. Then I went home, drank beer and water and Gatorade, fed Claire and Tom another meal Kathy had left in the freezer, and fell into bed. I was too tired to sleep, and lay awake for hours worrying.

Claire awakened me at 2:30 A.M. A cataclysmic storm was shaking the house. Rain fell in heavy layers; thunder boomed and lightning crackled. Jack and Doty cowered, and flashes illuminated Claire's face, pale in the night. Rain and wind beat a peach tree to the ground outside our door. The sheep would be huddled in a knot at Mossy Dell, heads lowered in the maelstrom, lambs against their mothers.

"It's okay," I told Claire, who held Jack, trembling on the couch. "We're safe. It's just a storm."

I listened to the rain's drumming and imagined what this meant for the pond and the naked earth across the road. In an hour, when the storm quieted, I returned to bed somberly. *Why did I start this project?* My worst fears had come true.

In the morning I sped into the farmyard. Our pond bulged from rising six inches, its new trickle tube submerged in muddy water. The dam had held. A bucket I'd left out contained four inches, rain that had fallen in only one hour, an incredible deluge. The wavy lines of debris across the gravel farmyard indicated it had been under water in the night as runoff from the hills and our neighbors' driveways along Snowden Road flooded the roadside ditches and raced toward our little pond. Beyond, I saw that the straw had saved the hillside.

I walked onto the dam and looked down. The spillway was destroyed, again a jagged ravine. Water, surging into the safety notch at the pond's lip, had torn loose fresh dirt; the slurry cascaded down the back of the dam, picking up speed, and tumbled across the lower pasture in the pounding rain. The flood stripped the topsoil from the center of our newly graded field. Between the dam and the fringe of trees along Lake Snowden lay a sheet of slick orange subsoil. Bulldozer tracks crisscrossed the mud, each imprint a rectangular puddle glinting in the sun.

The government technician said he'd never seen anything quite like what had happened. What to do? Daniel was gone and there was no getting him back. He helped me locate a couple of guys with nimble equipment for repairs. They came right away, and I bought them twenty tons of topsoil and twenty tons of rocks to slather over the storm's wounds. I canceled our Florida vacation and spent days spreading seed, old hay, and straw on the spillway's banks and in the waterway below. I paid the bills for the reconstruction—$1,100—and dreaded the invoice for the main job: four days of moving dirt, not the three Daniel had estimated. Kathy wouldn't be happy.

A ragged edge remained, a riddle I couldn't solve. A good man, hardworking, competent, had taken a life. His younger self, long gone, had been negligent and had left him stranded to relive for decade after decade what had happened on a dark road one night. I imagined that Daniel might have continued to pay the girl's family after, what, fifty years? It was the kind of blunder—one moment in a life—that I could now so clearly imagine making. Yet I couldn't fathom living with it. I thought I'd seem more broken. Maybe it didn't bother Daniel after so long, yet that was even harder to imagine. Surely even Daniel couldn't outrace guilt and sorrow.

You can't ask someone about such a thing. Doing so would be monstrous, selfishly seeking clues to the mystery the past has for any adult. Trying to comprehend the person you'd been, to grasp your own failures or sins. Probably his answer wouldn't help anyone else. He'd just suffered.

The memory of that terrible storm faded, and grass softened the contours of our brutal work. Years later, with the kids away at college and Mom dead and gone, I regretted backing out of our vacation. Sometimes I thought about Daniel and his awful burden. I'd remember his story forever but never plumb its awful depths. How does anyone atone?

All I know is that Daniel never billed us.

Something is always going awry, getting out of control, and otherwise cheating one's fantasies on a farm. Every growing season offers the potential for a disaster of biblical proportions. The example of how Dad adapted and kept going was what inspired me. He didn't expect life to be easy or fair—it hadn't ever been. He put his head down and got back to work. I remembered how, when he'd started his nursery, his first plants started dying: he tested his irrigation well and learned that it was almost as salty as seawater. I feared this meant the end of his latest dream. Instead he switched to growing native plants that had evolved to tolerate salinity.

As he worked long days in the sun and his nursery prospered, his health declined. His back problems had stabilized, but his heart, scarred from two major attacks, was weakening. After six years as a nurseryman, he decided to close another chapter. Not only would he go out of business, he would sell the latest world he'd created: the board-and-batten cedar house, the rows of plants in black plastic pots, the duck pond, Mom's guinea barn, the muscadine arbor.

"Some people just retire," I told him during a visit. "They enjoy doing things around a place." He said nothing, just shook his head, looked away— saying it all. He'd never putter, rest on his laurels, do nothing but relax—all of that a living death to him. He engaged with the world through work, and would increase his duties as a consultant for Pan American.

Still, I'd like to think my parents might have stayed there, entering a new phase of Dad's form of retirement, if a developer hadn't bought the pinewoods beside them and built a twenty-four-hour truck terminal. Wax myrtles, oaks, and slash pines couldn't shield the nursery from the glare of floodlights, or soften the noise of trucks grinding their gears at three o'clock in the morning. Dad got the developer to replace a woven-wire cattle fence with six feet of chain link, which only emphasized how much had changed. At the time, the farm's fate seemed quintessentially Floridian and especially cruel, but such gross transformations of landscape are commonplace.

And soon the truck terminal's owner bought Coral Tree Farm. As my parents buttoned up their place, the man prepared to move into the gracious house he'd driven them from, the house where, in a few short years, they'd made so many memories. The oak-shaded plot was where Dad had returned to the land—once again, one last time—and with a mighty will had resurrected his boyhood dream.

Mom had always accommodated Dad's sudden changes of locales and careers. Now she was ready to enact her own plan. She was displeased by the way Dad never consulted her. After his sale of Coral Tree Farm, she told him she was separating from him and would move to an apartment. He was dumbfounded. It had been thirty years since their last divorce.

"I want to be the captain of my ship for once," she said.

"I can't believe you're doing this at sixty-eight years old," he said.

"You just watch me," she replied. "Have I ever said I was going to do something I didn't?"

Mom had no trouble recalling this conversation for her children, but she seemed less able to explain her need to break away. My sister told her that her response to Coral Tree Farm's sale seemed excessive. But Mom needed to escape from Dad's darkness. "If I don't get out I'll die," she told Meg. "I have to save myself."

The next time I saw my parents, they were living apart—Mom in an apartment in Orlando, near Meg's home, and Dad in a one-bedroom condominium near the beach in Cape Canaveral. Mom had helped him find the condo and visited him there, though she never stayed over. "He's nervous as a cat when I'm there," she told me. I imagine this was because he feared criticism of his domestic skills. The place was spotless, of course. He didn't cook but ate all his meals out—"He can't boil water," Mom said—and he had a housekeeper.

Dad had retained his quality of exile. A visitor who'd been stranded when his world had flickered, dimmed, and died, he was apart from other people. I'd never had the sense of his moving through life with us. In Cape Canaveral, his apartness was heightened by the fact that he was in literal exile, banished by his Rosie to get his mind right one last time.

He'd started over as he always did, with a clean slate, retaining nothing from Coral Tree Farm. Not a photograph. Not a screwdriver. His pared-down existence was epitomized by the suburban electric weed whacker he bought for cutting the strip of lawn he was responsible for in front of his condo. He kept after that coarse swatch of turf with his string trimmer. When I visited him once, he buzzed the grass, a two-minute job, and the motor's whine attracted the attention of his neighbor. The unshaven young man ambled outside grinning and hailed Dad warmly: "Hi, Charlie!"

I was appalled and must've looked astonished. *Charlie?* This slob thought my father was like anyone else, an ordinary human, undignified—mortal.

Before he left for his missionary work in Guatemala, Mike Guthrie had found us an easygoing handyman, who'd appeared erratically over the summer to chip away at finishing the house, his schedule fitting our cash flow. And in August my mother loaned us $5,000, insisting that we at least finish the master bathroom instead of using the children's. Mom thought that Kathy, a busy mother with a demanding career, needed a more comfortable home.

We'd reduced our debt by half, and Kathy wanted to finish the rest of the house, including the upstairs and the wraparound front porch. Meg loaned us $15,000 to pay off the credit cards; now debt-free, at least officially, we qualified for a home equity loan from Athens's credit union. We paid Meg back immediately from the loan and had our carpenter redouble his efforts. The oak trim finally went up, beautifully amber with swirls in the grain, beneath our home's gracious nine-foot ceilings. We couldn't savor the milestone, just checked it off. It would be years before I would admire that wood, though our occasional dinner guests marveled at the simple, elegant Appalachian pattern, poised between refined and rustic, that the craftsman in Shade had picked.

Our house disaster receded as the hot, dry growing season of 1999 drew to an end. We'd moved from Indiana only three years before, but it felt like a lot longer. I'd been tested by dreadful Fred, by outrageous weather, by the chaos of lambing, by hated erosion. But mostly by my own stubbornness and mistakes. We'd overcome much, had prevailed, and could almost relax. At the press, we'd soon celebrate our reissuance of Louis Bromfield's daughter's 1962 memoir, *The Heritage*, about her father.

And finally, one afternoon I walked into the weeds covering the cornfield and found beneath rank Johnson grass and ragweed a thick growth of tender lespedeza, which—true to its reputation—had loved the droughty summer. Undeterred by the poor soil, the legumes were emerald and looked like fine young alfalfa. I called the Farm Service Agency's forage expert, who visited and expressed amazement at the success of the species I'd chosen.

One Saturday that September, Kathy and I waded into a pen of baaing ewe lambs inside the cavernous, dimly lit white barn on the Goss farm. The air felt humid with lambs and smelled yeasty with the amalgam of a sheep barn: mellow musk, tangy manure, bitter urine, sweet hay. Glen Fletcher had packed the animals so tightly they could hardly move, and our legs brushed soft lamb coats, beneath which we felt dense bodies. One of these bruisers could take out a knee. But the lambs were calmed by being so crowded. When we stopped, they looked up, unbearably cute with their bright eyes and their mouths pursed in the beginning of the faint smile that adult sheep seem to carry on their gently upturned lips.

"I want to give these guys noogies," I said. "They're beautiful, Glen."

"*Kathy* picks," Glen said. He backed against a whitewashed wall, assuming a hands-off air, bemused. His muscular arms for once idle, he tried to set his normally smiling bearded face into a mask. His countenance was layered archetypes: the cat that ate the canary, a poker player holding a good hand, a professor letting the student make her own mistakes.

I hung back too, as if I also knew what I was doing, the farmer who'd turned a manly duty over to the little lady. Glen liked me, but I knew he could see my pose. Kathy, a bit perplexed, stood with her hands on her hips and looked down. I felt a surge of excitement—Glen had given us the pick of the farm's ewe-lamb crop!

He'd always skimmed the best-looking lambs for the home flock. Normally he'd keep all these lambs, to grow the flock and to replace aged ewes being culled for bad teeth or damaged udders. Now he'd reached the Gosses' goal of one hundred Katahdin ewes. And sadly, the future of the flock was in doubt. Hank had died the previous winter from his cancer, and Mary had moved into town.

"Mary says we're keeping the flock going," Glen had said over the phone. "The kids want her to visit the sheep. I guess we're taking things one day at a time. But I can let some lambs go. Is Kathy coming?"

Glen's uncertainty had made me greedy. After my experiences with poor mothering, crazy behavior, birthing problems, and highly variable size and appearance, I was obsessed with getting better genetics. Kathy was my secret weapon. Glen liked her, and I was sure he'd sell me better animals with her along.

"You *have* to come this time," I'd told her.

Although we now had twenty ewes of our own, eleven surviving originals and nine ewe lambs, I wanted to expand faster than our flock's rate of reproduction. Mossy Dell grew enough forage for forty ewes, plus there was the crop field I'd plant to grasses next year: as a pasture it could support another sixty ewes, and our hilltop could graze fifty ewes as soon as I added woven wire and more posts to Fred's barbed-wire fences.

"That one," Kathy said, pointing to a flashy brown lamb with a white face. Glen sprang forward and the group moved away like a school of fish. But he'd both caused their movement and gotten ahead of the chosen lamb, and he collared her neck in the crook of his arm and scooted her into an adjoining stall.

Thrilled with the lamb, I still wished Glen would help us. He was an experienced judge of how a sheep would mature, and I craved bigger, stockier, prettier ewes. Cognizant of my desire, he'd put Kathy in charge for his own reasons, which seemed born of respect for her, chivalrous flirtation, and a desire, I suspected, to discipline my rapacity for sheep.

"Her," she said, pointing now to a big, dazzlingly white lamb with a pink muzzle. Glen caught the animal in two quick moves, hustled her out.

"That one," Kathy said, indicating a chunky, low-slung white lamb with a few tan spots on her belly, red hocks, and a spray of black freckles across her face. I patted my checkbook.

The Goss ewes had lambed in this barn in January and February, and Glen had fed them and their offspring—these lambs—oats and alfalfa hay. Then he'd grazed them all summer. The lambs looked huge, especially compared to ours, three months younger and reared on grass. I wanted ewes that lambed on pasture, but they were hard to find, and these were such a deal. I knew Mike Guthrie would be disgusted with me for buying future brood ewes based on size and beauty. I could hear his refrain: "It doesn't matter how they *look*, Richard. It's how they *perform*. Big ewes just eat more."

I knew. And I'd read enough to know better. This wasn't the approved way to pick breeding stock, though the usual one. As at Jo's, no data guided us: no lamb weights taken at weaning to indicate growth rates, no records telling us whether these were singles, twins, or triplets. Not even ear tags to indicate parentage. In the waning days of the Goss farm, home to sheep and shepherds for three generations, we sorted blindly through what might be the old farm's last lamb crop.

And I couldn't stop grinning. These were going to be beautiful sheep.

"That's an even dozen," Glen said with finality. I felt a mix of relief and thwarted avarice. I was hoping for fifteen, maybe twenty. I hadn't wanted to dicker, offered this opportunity, and hadn't even asked about price.

"What do I owe you?"

"Oh, I don't know," Glen said. "What's the market price right now?"

"I haven't sent any sheep to market."

"Let's say sixty each. Is that fair?"

"It's a bargain," I said, writing the check. An incredible bargain. At last. And in one stroke we'd increased our flock's size by a third.

"I might need to borrow George back for breeding," Glen said.

"He's yours. I don't breed until Thanksgiving."

"I'll come get him and bring him back in time for you."

Feeling a rush of gratitude, I blurted, "I've sent Laura Fortmeyer a deposit on a ram lamb and am going out in October. Would you like to come along, maybe get a new ram yourself?"

"Great," he said. "*Kansas.* I'm always ready for a road trip."

Glen and I were so different—he was pure country—yet in the speed of his assent I sensed his reciprocal feeling. We shared a kinship, that continent below surface differences.

Over at Mossy Dell, Flower watched the flock. She was an escape artist and I'd given up trying to confine her to the electrified paddock with the sheep, so she rested in the shady field borders in the day. She visited the gravel farmyard at will, and at night patrolled the farm's perimeter against coyotes.

One day when I pulled into the farmyard I saw her playing like a puppy with the black dog, ignoring her flock. She'd grown up with sheep and was bonded to them, her hoofed pack, but was lonely for her own kind. The stray slunk away as I stared. I sensed it was a menace to our flock, and I worried that Flower wouldn't stop an attack.

The blameless thing to do was to drive down the road and talk to Johnny and his wife, Penny, and ask them to keep their dog on their own property. I didn't think that would work. Johnny was one of Fred's hunting buddies, and Penny was related to Dolores. I suspected they'd tell Fred about my complaint and deepen the Paine clan's animosity toward me. In July our mailbox had been beaten in, and I suspected Fred's grown son, Shane, who lived nearby and helped Fred plant and harvest his properties, including the field beside our house. I'd met Shane only once, briefly, before the house debacle, and he'd been unfriendly. Now he violated rural etiquette by refusing to wave when he saw me on the road; maybe he hated us for publicly gutting his boyhood home.

"How do you handle neighbors' dogs?" I asked a cattleman on my next pasture walk.

"They don't usually bother cows much," said Ben. "I used to keep sheep, too. Some city people moved in next door and their dog started harassing my flock. I told them, but they ignored me, and one day I shot him."

The dog, a Husky, was chasing Ben's ewes, driving them with a bloody mouth. There was wool stuck between its teeth—Ben could show the owners that proof—but when he knocked on their door and told them what he'd done, they went berserk. They didn't stroll outside to see the bleeding, trembling ewes—their udders ripped open, divots of flesh torn from flanks and necks—that Ben also had to shoot. "They called me a murderer in front of their children," he said. "They wrote letters to the editor calling me a heartless killer, and threatened to sue me."

He was acting within his historic right as a shepherd, as well as lawfully, and claimed recompense from a county fund for livestock killed by dogs. (He also could have demanded payment from the dog's owners.) But he was so upset by the couple's outrage that he sold his flock. A hundred years ago, when there was an intact rural culture in Ohio, Ben's first warning might have resulted in the owner dragging his own dog off the porch and shooting it himself. Now even America's rural dogs had morphed from working helpers to idle companions to sacred creatures.

I couldn't countenance a dog attack on our ewes. I could imagine it too well from hearing shepherds' stories and seeing our chickens massacred. I loved dogs—as a boy, I'd stood weeping and pleading for one in front of my embarrassed father—but I decided that having sheep killed and maimed would be far worse to bear than would shooting the black dog.

And yet killing rogue dogs, once understood as necessary when farms were diversified, each with several livestock species to protect, had become a shameful secret. Shepherds on Sheep-L often cited the pragmatic, contemporary Three S Rule: shoot, shovel, shut up.

I started carrying my rifle in my truck's cab.

To get an early start on our trip to Kansas, I drove west that Thursday afternoon to Dayton, where I'd bed down on Glen's couch. The sun was low in a yawning blue sky, bright and fathomless, and the roadside trees' pale golden foliage seemed to catch the clear light from beneath, so that every fluttering leaf glowed. The air was cool and dry from a ridge of high pressure that stretched from the Appalachians to the Rockies.

I didn't find Glen at his homestead—a compact white frame house trimmed in brown, a shiny red barn trimmed in white. I admired his forty-by-forty vegetable garden, still pristine this late in the season. When Glen had come into their lives twenty years before, the Gosses had helped him get back on his feet with this place to live at the far end of their farm. He was newly divorced and broke, with two small children to rear alone. He'd been down in Atlanta, making good money working construction—and the brothers who owned the company promoted him to foreman—but his wife demanded they return. Then she left him anyway. He still exchanged his labor on the farm for rent, and he'd remodeled the house and erected the barn.

His world looked timeless, yet wasn't. On my way to the big white barn I drove past the vacant Goss farmhouse. I was glad he was going with me, but wondered how he felt about buying a ram in the name of an elderly widow whose grown children indulged her the sheep for sentimental reasons.

I found him behind the barn, filling a water trough with a garden hose. "There he is," Glen said. "We better get supper and hit the sack. Gotta leave *early.*"

On Friday we crossed Ohio, Indiana, and Missouri, and entered Kansas. We laughed and talked nonstop in my truck. About what, I can't remember, though I almost lost my voice from unaccustomed use. We were just two friends, barreling west in a white Ford pickup, laughing under the bluest sky.

The next morning, after a quick breakfast at our motel, we drove to the Fortmeyer farm and arrived midmorning. The place had the no-frills appearance of working grass farms: a spare white farmhouse, a modest machinery shed, an unpainted wooden hay barn. Lonely acres of grass rolled away to distant breaks of ragged trees. Laura and Doug Fortmeyer, George's breeders, had met at Heifer Project International in Arkansas, where she managed the sheep flock in the 1980s. After Heifer disbanded its animal holdings, the couple brought about ninety Katahdin ewes here, to the 190-acre farm begun by Doug's grandparents. Now Laura and Doug were full-time producers of hay, freezer lambs, and breeding stock. Laura also was operations manager for the Katahdin society. I was interested in their bloodline not only because of their long tenure as breeders, but because they were shepherds who lambed outside on spring pasture. They selected for mothering ability under those more natural but demanding conditions; for years they had culled ewes that rejected lambs or couldn't tend them while foraging with the flock in big, busy pastures.

"Hi, Richard," Laura said. "You haven't been here before, have you? We haven't met?" Her voice was big on the phone, so I was surprised to meet a petite woman. She had dark hair and blue eyes, and was holding a clipboard. Numbers in columns, hand-drawn with a pencil, covered her yellow legal pad.

"No, this is my first visit," I said. "I've just called you a lot for advice."

She walked with me to the barn, Glen and Doug ambling behind. In the pen of ram lambs, a large sleek white one stood out.

"That's the one I pulled out for you," Laura said.

"I want him."

"You like him?"

"Oh gosh, yes."

"I'll show you his mother. He's a twin and grew well. Here's his weights. We had him reserved for a woman who's had a deposit in for a while, but she had a heart attack."

I thought Glen looked jealous as Laura caught my lamb. But I'd sent a deposit on a $300 animal, and he was buying a $250 ram, no deposit. Glen entered the pen and peered at the lambs.

"I like that one," he said, nodding to another white lamb.

"He was born a triplet but we pulled one and raised him as a twin," Laura said.

After Glen caught the lamb we looked at the Fortmeyers' ewes and rams, had lunch in their kitchen—lamb stew and lamb summer sausage—and then got back on the road, our two white ram lambs under my blue topper. In late afternoon our lambs folded their legs and rested against each other, shoulder to shoulder, buddies. The low sun seemed to push us as it lit our path, and my truck followed its own dark shadow toward Ohio. We ate barbeque at a truck stop somewhere in western Illinois. Afterward we were talking about whether to stop for the night, Glen driving, when I fell asleep. When I awoke it was one o'clock in the morning.

"Where are we?" I asked. I wanted to be home. And I looked forward to next weekend, when Ellen Bromfield Geld and her husband, Carson, would visit Athens from their ranch in Brazil. She'd sign my copy of her memoir. They would see the sheep at Mossy Dell and laugh at Flower, who ran from strangers, as timid as a sheep; they'd dine at our house on the hilltop.

"Almost to Dayton," Glen said. "You really conked out. Must be the excitement of getting that big ram."

I sold my first breeding stock that fall: the little brown ram lamb that Kathy had saved from my neutering, Brownie's son.

On a damp Friday morning in early November, Bryce Chatsworth drove into Mossy Dell's farmyard from West Virginia in a dented white pickup with stock racks. Black letters on the driver's door said, "Chatsworth Stock Farm, Bayard, WV."

"I grew up not far from here," he said after we shook hands. "My dad moved to Ohio for work, and we farmed on the side. I went back ten years ago and took over my grandparents' farm." He said it was high elevation, against the border of western Maryland, and all bluegrass—like upstate New York, with snow to match.

We walked uphill to Massey's shed. I had tried to discourage Bryce on the phone when he'd called in response to my ad.

"I do still have the ram lamb," I'd said. "He's brown." I knew most commercial shepherds wanted white sheep, and this man sounded serious, all business.

"You want $150 for him?"

"Yes," I'd said. Then warned, "He's not real big." My new ram lamb from Kansas made Brownie's son look fey to me, hardly masculine. Fine for a hobbyist who wanted color, maybe.

"I don't care about that," he'd said. "I need shedding genes to get rid of wool. I'm losing money on it. Do you have any ewes you want to sell?"

Now he stood in Massey's shed and looked at the little brown lamb. "I'll take him," he said. "These are the yearlings? Did they both lamb?" In the next pen were The Bad Mother and another ewe that had upset me.

"Yes, they both raised lambs. They were kind of . . . flighty."

"Yearlings. Maybe they'll calm down. How are their udders?"

"Fine, as far as I know."

"I'll check them. Is this the old ram you mentioned?"

George rested on his belly in his stall, ignoring us. Glen had brought him back limping, George having injured his shoulder—or maybe he was just showing his age. I'd decided to breed everything to my new ram lamb and thought I should sell George if I were going to idle him for a year. I feared having to ship him. And I couldn't imagine having him turned into sausage for our table. Yet I was sentimental about him, and now regretted throwing him into the ad with Brownie's son. I felt like such an amateur with this pro breathing down my neck, a man who'd load up my entire fretted-over flock if I'd let him. He'd driven from West Virginia for three lousy sheep.

"A woman called and said she wants him," I said.

"Did she send a deposit?"

"No."

He stared at me. "You'll learn to sell to the person who shows up with money, wanting to buy," Bryce said. "Now let's load these sheep, and each of us decide to be happy."

So I'd made my first sale, a drop in the bucket, despite myself. And soon our new young ram—whom I called Kansas, naturally—swaggered among our thirty ewes with a free-and-easy macho stride, busy making our second lamb crop.

CHAPTER TWELVE

Brownie had twins again. Couldn't work them with a hurt wrist.
—Lambing Notebook, 2000

I'D FIRST NOTICED CREAM'S PROBLEM IN LATE GESTATION, JUST before our second lambing began in April 2000, when her anus bulged like a bright red ball. Now, several days after she'd lambed, her rectum protruded and hung down like the trunk of a baby elephant. She'd suffered a rectal prolapse. After last year's lambing, I knew to expect anything, but this was a rare malady.

Sometimes the rectum prolapses in overfed feedlot lambs, especially those with closely docked tails and respiratory ills that cause hard coughing. But it's unheard of in mature sheep. And Cream wasn't even docked. Like all our ewes she retained her long tail and the muscles at its base that lifted it away when she pooped and peed. Tails on sheep look strange to shepherds—"Like the ewes are lions or something," Mike Guthrie said—but I'd seen the appendages switch flies and flag down a passing ram. Even if Cream had been docked, the pressure of pregnancy was over; she'd had a big single.

I tempted her into Massey's shed by rattling grain in the bottom of a bucket. I caught her, haltered and tied her, and shoved the rubbery bowel back inside with my fingers.

Like our pet sheep, Red, Cream had personality. A rangy ewe with a prominent Roman nose, Cream hated Flower and would sneak up on the dog and butt her. This was so mean and devious, and Flower's reaction so meek and abashed, that I couldn't help but laugh every time I heard the *whump* of Cream's ambush and Flower's yelp.

Red also had had a single. I knew this wasn't desirable, for two-year-olds to raise one lamb, but Red had raised twins as a yearling. "She's making up," Glen Fletcher said. Mike disagreed: "She should've twinned again, at least. Mine that twin as yearlings triple as two-year-olds." But once again, our sheep hadn't read the book.

A ewe that had tripled was neglecting one of her lambs. She didn't overtly reject it, but she ignored it as its stronger siblings pushed it away when it tried to nurse. One dewy morning I found the soaked lamb lying alone in the grass, chilled and weak, and brought it home for Claire and Kathy to raise on a bottle. Persephone, as Claire called her, nursed every four hours; Kathy took the midnight and four A.M. feedings.

One of the new yearlings we'd bought from Glen distinguished herself by twinning and taking good care of both lambs. The chunky little ewe with the black spots on her nose—Freckles, I dubbed her—kept her babies away from the flock. Her family moved together, shoulder to shoulder, as if welded. "Great mother!" I wrote in my lambing notebook, and gave her a perfect maternal score of four, the rarified rank of Red and The Good Mother. More maternal rightness.

Meanwhile, Cream struggled. In the week after assisting her with her lower bowel, I had to do it twice more. And then on the weekend I saw it again, swinging below her tail. The exposed organ was now swollen and raw, with pale tissue sloughing off. I tempted her near Massey's shed with corn again, shooed her inside, and went to work. This was becoming almost routine—her lamb stood by patiently—but I feared infection and knew flies would become a threat as the weather warmed. I used lots of antiseptic lube this time as I worked the protuberance back inside. "That's it," I told her. "This is the fourth time. If it falls out again, you're on your own."

Finally, as if she'd listened, Cream kept her rectum inside where it belonged.

By late April I was wearing a brace on my right wrist, sprained moving Electronet. I worried constantly about not getting any permanent fencing done for our growing flock, and one day my neck broke out in prickly welts. "That's shingles," my doctor said. The reemergence of the chicken pox virus is serious if untreated. He wrote prescriptions for an antiviral medicine and an antibiotic salve.

I googled shingles: brought on by stress.

Then, one weekday in May, Fred demolished with his yellow bulldozer the fence between Mossy Dell's northern pasture and the back corner of Johnny's property. Johnny, he of the roving dogs, had mentioned he was planning to run cattle. "I'll need to fix up the line fence this summer," he'd said. "It won't hold cows." Now, without talking to me, he'd asked Fred to clear the line. Fred's work looked brutal, with ruts in the earth, and gouged trees with branches hanging broken. He'd snapped off Massey's old locust fence posts and pushed them and other debris into Johnny's pasture. As I stood looking at the mess, Johnny came striding back, gripping a Budweiser. He told me his cattle would arrive the next week.

Replacing the shared fence line became a desperate weekend's ordeal—for me. For Johnny it was an occasion for visitors and more beer. I'd bought new posts and he'd gotten a roll of barbed wire, and as we drove the posts that Saturday his friends showed up, Shane Paine among them. They didn't acknowledge me, or offer to help, but they joshed Johnny and bore witness. Their women carried drooling babies that gaped at my green tractor and red pounder.

Even by the standards of Appalachia, Johnny was casual. Along the road, he and Penny had planted a vegetable garden, now overtaken by weeds. They kept some ducks, and when I'd given them eight baby chicks in April—one of our hens had hatched too many to care for—they'd let them die somehow. Maybe their dogs had gulped them down. Johnny affected a southern drawl, and a Rebel battle flag flew from a pole in front of his trailer. His ethos seemed to have sprung from the *Mother Earth News*, *Guns & Ammo*, and *Iron Horse*. Sometimes he helped Fred and Shane harvest crops or build houses, and he commuted to Columbus to work when contractors were hiring laborers. Before dawn, I'd hear his muffler-less pickup roar to life up Ridge Road. Today he wore cracked leather work boots with his canvas Carhartts, and a beer belly drew taut his orange Harley Davidson T-shirt.

"Mister Gilbert?" Johnny offered me a Budweiser at the end of the day from the cooler on the tailgate of his Chevy. He held it toward me, and a friendly smile parted his scruffy brown beard. I'd been worried when I saw the cooler. Drinking beer while operating a post pounder was the stupidest thing imaginable—I could see the headlines—but Johnny had waited. "Thanks, Johnny," I said. The truth was, I envied his insular world and uncouth friends. He had friends. And his buddies had time to drive over on a Saturday and hang out, enjoying the mild weather, watching other people work. All I did anymore was work, either at the press or the farm, and I couldn't keep up anywhere. My coworkers thought I was crazy, I could tell, although I didn't share a tenth of what was going on.

As I left Mossy Dell that evening, Sam Norton flagged me down on Snowden Road. "I saw how Johnny and Penny built their first fences," Sam told me. "She pulled the wire with her *hands*."

I pictured Penny bending forward to choke up on a strand, then rearing back to hold it as Johnny sauntered over to hammer the staples. "You can't get barbed wire tight that way," I said.

"Oh it's the sorriest thing you'll ever see!" Sam cried. "Take a look when you go past, if you can see it through the pines. That's another thing. He planted those trees and didn't water them *once* all summer. Why, if you or I would've done that, *every single last one* would've burned up!"

"He's laid back," I said.

"Lazy hippie is what he is."

Sunday afternoon we stretched the barbed wire using my come-along, a ratchet stretcher I'd inherited from Dad. I liked squeezing the textured blue plastic that wrapped its steel handle, giving a crank, and feeling strong as a strand tugged and came taut. I wasn't sure how hard to pull barbed wire. "I don't know either," Johnny admitted. "Just jerk 'er good and tight." As I knelt on the scar of raw dirt, Johnny's skinny tan pit-bull cross sidled up behind me, sniffed the soles of my boots, and then sprinted downhill toward the trailer.

Several times recently I'd spooked the dog out of our driveway, and usually saw his black companion staring at me from up the road. Once I'd thrown gravel at the tawny dog, and twice I'd aborted trips to town rather than risk leaving with him lingering near. The second time I'd been with Kathy and the kids, going out to dinner, and had gotten out of the van and sent them on. I resented Johnny's unknowing usurpation of my precious time. The only question was whether his dogs would hit the chickens on the hilltop first, or the sheep at Mossy Dell.

"Johnny," I said, rising stiffly, "your yellow dog has been coming up our driveway. I'm afraid he'll kill our chickens."

"Chopper wouldn't hurt anything," he said. "You should see how loving he is to us."

I was astounded. The man was a predator himself, an avid hunter, part of Fred's well-armed posse, so he seemed willfully ignorant. *Chopper.* The name said it all. And yet I knew people couldn't imagine their pets as killers; I'd never forget my own shock at seeing Doty, the meekest of collies, shake a groundhog to death in her teeth back in Indiana.

"I'm concerned about my sheep too," I said. "I'm going to graze them around our house soon." I didn't mention the black dog's visits to Mossy Dell, my true fear.

"I don't think he'll chase them."

"He'll chase deer, won't he? If he'll chase deer, he'll chase sheep."

"Oh," Johnny said, gesturing with his beer, "just shoot my dogs if they get in your sheep."

He was serious. It was quite an offer, an affirmation of his rural heritage. But it wasn't good enough. I couldn't wait until an attack happened. I'd get there for the aftermath, our ewes bleeding, our pretty new ram dead or his testicles torn off. And what would happen when Johnny told Fred and Shane that I'd killed his dogs? At the least, we'd never have a functioning mailbox. And terrible things might be visited upon our livestock in the night.

I drove out of Mossy Dell's farmyard aching—I'd strained my creaky lower back—and irritated. My neck prickled where the shingles had subsided. I wanted to get home, soak in the tub, drink my own beer. But ahead of me I saw Sam, standing in his driveway, home from church and his weekly restaurant meal. Three of our dome-shaped gray guineas raced across his yard; even from a distance I could see their leathery faces, as white as if caked with pancake makeup. They wouldn't hurt Sam's garden or flower beds, since they never scratched up the dirt like chickens, just pecked at bugs, but I knew their untidy trespassing must irk my fastidious neighbor. I was guilty of what I'd just condemned Johnny for, the rural sin of failing to control my animals.

"I see some of our guineas are over here. You can shoot them, Sam."

"They're not bothering anything."

"It's fine by me if you pick them off. Those that leave our land are the ones we don't want."

"How's that fencing going back there?"

"It's gone. It went. It's done."

"Fred kinda messed up the line, didn't he?" He looked at me slyly. He'd probably watched Fred work, or had sneaked back to witness the aftermath.

"Don't get me started."

"I'll show you my stoop sometime," Sam said.

"Your stoop?"

"Me and Erma hired him to dig the foundation for our sun porch. Shane poured the concrete for the stoop too high. The door opens out—it was cemented shut! I told Fred, and he sent Shane back with a chisel to chip away the concrete so the door could swing. It was *awful*."

I burst out laughing. Telling his tale, Sam had shaken his crown of flossy hair and snapped off his words. But a sparkle in his blue eyes showed that he saw the humor.

Evidently, keeping track of Fred was a neighborhood sport and something of a necessity.

One Saturday that June, after drenching the flock with worm medicine and moving them to fresh grass, I sat fretting after dinner. I'd felt weak all day and now was exhausted.

Last week I'd found two husky ewe lambs dead. They'd lain down under the shade of a hickory and never got up. There wasn't a mark on them: internal predators, worms, had killed them, bled them to death. Before I could drench, another three lambs died. Losing five nice animals was the price for buying the Goss lambs the previous summer and failing to clean out their parasites, as I'd done with our first sheep. Glen had helped me deworm them as we'd loaded them on my truck, but the Goss farm's nematodes probably were resistant because Glen had used the same product for years. Now that the pastures at Mossy Dell were dirty, I'd have to medicate lambs every three weeks.

I could see from my seat at our table a section of Fred's barbed wire on the hill that rose past the little pond in the south pasture. With my day job, it was difficult to get any projects done after chores. And building fences took forever. Almost as many acres of grass were going ungrazed around our house as were in our vast unfenced former cornfield across the road. This bugged Fred, or maybe he was just trying to bug me, because when I'd run into him one day at the Albany post office he'd asked why I wasn't pasturing his old homestead. Reinforcing his barbed wire was the fastest way to get more clean pasture. *But it's already June*, I thought. *I'll never make it.*

Suddenly my chest felt like it was being crushed. I fought to breathe. I waited for the pain to pass, but it didn't. I rose and got our family medical guide and read the entry on heart attacks. Kathy, cleaning up the kitchen, glanced at me as I reshelved the heavy book. I walked past her into our bedroom, put on a clean shirt and pants, got a pair of socks from my maple dresser, and slammed its drawer. Fear and self-pity fueled a wave of anger. *I'm dying and Kathy's oblivious.* I sat on the bed to put on my town shoes. Again I tried for a deep breath: impossible with the tightness in my chest.

"What's going on?" Kathy asked. She stood in the doorway.

"Nothing."

"You've been snapping at me today. What's wrong?"

"I hurt. It's like there's an elephant on my chest. Maybe you should take me to the emergency room. I'd drive myself but—"

"Kids," she yelled, "I'm taking your dad to the hospital. I'll call when I find out what's going on."

Heart attacks had laid low both my father and his first son, my step-brother Chuck. My mother had undergone bypass surgery. Everyone in my family assumed it was only a matter of time before I or one of my younger brothers was felled.

The doctor approved doses of morphine as I lay on a gurney and arched my back in pain. I held Kathy's hand and saw the nurses shake their heads at my resistance to the painkiller. "That doesn't even touch it," I said, looking into Kathy's worried face. But with the pain, I felt relief, the passive, self-centered acceptance of a patient, the surrender to something that dwarfed trivial daily struggles. Life—and death—were so much bigger than my plans. And now I didn't have to worry about the sheep.

Periodically Kathy left to call her youngest sister, Karri, a physician who'd returned to the Krendl farm and had a medical practice in northwestern Ohio. Kathy would return and I'd hear her ask the doctor questions. He furrowed his brow at my symptoms, which, other than chest pain, apparently didn't fit a heart attack. Finally he requested a helicopter to fly me to a hospital in Columbus, but the chopper wasn't available, so an ambulance would race me two hours north.

"Do you want to go with him?" a nurse asked Kathy.

"No, I can't," she said. "Our kids." I knew she also had our bottle lamb to feed. And now the ewes to check.

"The sheep are fine in the south pasture for another two weeks," I told her. I added, "If you need to sell the flock, call Laura for advice."

"I don't want to hear this."

"You have to listen. Ask Sam for help. There'll be lots of grass on the north side, but you'll have to set up fence. Give them the whole field and let them wander." I was still bracing against the pain but felt calm, unafraid, and soberly dramatic. "Call Mom," I added.

At the big city hospital, where I was kept for two days, doctors performed a heart catheterization and ran other tests. Finally they diagnosed pericarditis, a rare inflammation of the sack surrounding the heart. Symptoms include pain, anxiety, and breathing difficulty. The ailment can be excruciating, but is caused by a mere virus, in my case probably by the chicken pox resurgence that underlay my shingles outbreak.

Released from the high-tech cardiac ward, I was told to take Advil.

The July sun beat down. I rode my John Deere, and Sam Norton held the posts for my hydraulic piston to pound. Sam, compact and wiry, wore his usual uniform: pants and shirt of blue-green twill, leather work boots. We were working on a slope at the hilltop farm. Across the road the ex-cornfield needed mowing to stop the weeds from setting seed—everywhere I looked was work. Which was why, after my heart incident, at my mother's and Kathy's urging, I'd hired Sam. Kathy and I remained on a tight budget, but obviously I'd set more in motion than I could handle alone, and doctors' bills were more costly than any hired labor.

In a workplace, there's room for many gifts, and yet the vision persists of the farm as a small ship guided and crewed by, at most, a couple. In this case, the farm was my project, not Kathy's, though she pitched in all the time. Sometimes I wished for even more from her; I'd catch myself fantasizing that farming was her dream too. What a team we'd make with her confident energy and my obsession with details, my "love of subjects," as my father had once put it. Yet she supported me as much as any mate could. In fact, it now seemed she'd given me enough rope to hang myself, and sometimes I secretly wished she'd lasso me and slow me down. If Kathy had a fault, other than being always too busy herself—the dark side of her strength, seeing opportunity—maybe it was her impulse to try to make me, an anxious dreamer, happy. I hadn't just married a horse, as Termite said back in Indiana, and not just a woman with beauty and brains, but someone truly giving and generous.

So now Sam, a hired helper. With him aboard, I felt myself slide a notch closer to wholesome reality: I needed help, we had some money again, and Sam hungered for farm work. Even so, when I reentered the farm from my office's digital world, nothing seemed more impossibly slow and exhausting than building fences. And today, waves of weakness swept me. Just steering the tractor felt effortful, as if I might slide like Jell-O from its vinyl seat. I talked with Sam while we worked, partly as a way of slowing him down. Sam moved fast, on legs that were noticeably bowed. Even in my weakened state we were getting a lot done—I saw that two men can accomplish three times as much as one man working alone—but my lack of energy was starting to scare me.

"I've got to knock off," I finally told him. I'd wanted to keep going—it was Saturday, hours before dinner—because Sam wouldn't work on Sundays.

"Are you sick?"

"Weak."

"We've got enough posts in to stretch another section. I could do that Monday."

I hadn't let Sam work alone yet on something I was that particular about. Besides, stretching wire was the fun part of fencing. Yet it was tempting to think of progress being made while I was at work. And Sam had built more fence than I ever would. He'd grown up on a farm on Baker Road, only a mile from us. As a boy, Sam had driven workhorses in the fields and tended milk cows, hogs, chickens, and even a flock of ewes. (But he'd never eaten a lamb chop in his life: sheep weren't for home consumption.)

"Okay," I said. "Stretch that run and tie off the wire like we've been doing."

When I got home Monday afternoon, I rushed over to the fence. He'd pulled the woven wire so tight that I felt certain the posts would rip from the soil in winter's contractions. I'd read about how to use crimps in the woven wire to gauge a stretch, and this was too tight, the once-springy wire like welded steel. I'd have to pry out staples to lessen the strain and add extra bracing at the corners when Sam was at church and unable to spy on me. He surely knew by then how I wanted the wire stretched; his act showed how glaringly at odds was the way we approached the world. I'd studied fencing in books, searched glossy catalogs for the latest New Zealand products, and surfed the Internet. Although he was deferential toward me, when left to his own devices Sam would never follow my researched methods. He'd learned how to fence from his father, and collected tips other old-timers passed along. And overdoing things was how Sam expressed *his* anxiety.

He caught me that evening as I raced toward Mossy Dell for chores. I'd accelerated as I took the jog onto Snowden Road, but couldn't pretend I didn't see him running from his shop waving at me.

"I got that fence tight as a fiddle string," he bragged.

One evening a few weeks later, I pulled into Mossy Dell's farmyard and saw the black dog on the north bank, near the entrance to the pasture. An easy broadside. I reached for my rifle on the back seat, put the .22 to my shoulder, and shot. I was aiming behind her shoulder, for her heart, but couldn't see where the bullet hit. She flinched and fell over, as if in slow motion, and lay stretched out on the ground, motionless and silent. I approached her and saw her eyes—looking forward, into the distance—and I shot behind her ear and she jerked and died.

I grabbed the dog's rear hocks and dragged her body toward the fringe of woods around Lake Snowden. The buzzards would get her. I trembled and stumbled, unsteady from adrenaline. It hadn't been so hard, after all, to kill her. I'd discover, though, that I'd never forget her accepting eyes or how the light in them faded.

That night I didn't sleep, but pulled Kathy tight.

Apparently, Johnny told Sam his black dog was missing.

"You didn't see it, did you, the one he called Midnight?" Sam asked, looking intently and suspiciously into my eyes. He'd ambushed me again on Snowden Road. I wondered if he'd heard my gunshots or if he'd merely concluded that I likely had done it.

"No," I said. "I haven't."

A lousy liar, I was sure he saw right through me. My obvious guilt and his inquisition annoyed me. I knew he'd approve of the killing and just wanted to be in on a secret. But I didn't trust him enough to admit I'd shot a neighbor's dog. Sam so liked being favored with a dab of news that he might puff up his chest and hint to someone—even to Johnny—that Richard Gilbert up on the hill took care of that dog.

Sam's own practice was to kill an array of creatures. "Dirty, filthy things," he'd say about groundhogs, blackbirds, and even svelte deer—anything that threatened his lawn or garden or disturbed his peace of mind as he watched his bird feeder. I'd hear gunshots from his house across the road as another varmint bit the dust. This was the rural way, especially toward predators, which my neighbors killed with gusto. The previous winter, when I'd let a young man deer-hunt on the edge of the old cornfield, he bragged that he'd killed a coyote. I'd warned him specifically not to do so, and told him what a wildlife biologist told me: killing an innocent coyote just brings in others that might not confine their hunting to rabbits. But like Sam, shooting a wild hunter, a competitor, had made him feel righteous. Like he'd set things a little more in order.

Sam was terrified of snakes and had made the mistake of telling everyone about his hysterical response the time a serpent dropped from a tree into his lap as he mowed a pasture. People used that phobia, assisted by rubber snakes and even garden hoses in the grass, to induce especially oversized reactions. At Mossy Dell he would try to kill any snakes we came across, over my protests. "God put them here for some reason, Sam," I finally told him. "And though we might not understand His purpose, we should respect His wishes." I invoked the deity to frame my argument in a way Sam would find compelling. I'm sure it didn't work. He was confident his purpose encompassed smiting any creature he viewed as a scourge.

I'd seen a rat in the barn, but hadn't mentioned it to Sam because I dreaded his excited response—which might involve us taking everything in the barn apart—and because I thought I'd catch a snake and turn it loose inside to do biological battle against rodents. It amused me to think how apoplectic Sam would be over this idea. If the barn were at Mossy Dell—that is, on a farm that had been loved—several snakes and an owl would be living in it. But Fred had scoured the hilltop clean of biodiversity with his tanks of toxic chemicals, his heavy equipment, and his own gunshots.

And I realized my notion of natural rodent control was a fantasy one afternoon when I had the chance to catch a rat snake and carry it home, but couldn't do it. I stood a few feet from the reptile, a glossy three-footer lying on a trail around Lake Snowden, and thought I could grab it without getting bitten. But as I imagined carrying it to my truck, and then driving home with it thrashing in my lap and curling around my arm, I balked. I had my own snake issues, which I never confessed to Sam for fear of reinforcing his merciless practice.

A few days after I encountered the snake on my hike, a farmer delivering round bales of hay to our barnyard told me there was a big snake at the foot of our driveway. The weather had cooled, a taste of fall was in the air, and reptiles must've been on the move. I ran down there, thinking I could swallow my repugnance and carry it the short distance to the barn. But the snake already was dead, a driver having turned his tires to squash the creature.

On a Saturday afternoon in late September, Kathy left to run the kids into Athens for joint sleepovers. I climbed in my truck to go check on the sheep. The day had turned breezy, and the leaves on the red maples in front of our house rustled in the mild air. From the hilltop I could see a mile down Ridge Road to the south. The roadside cornfields had browned, a warm khaki, and late-planted soybeans were lime green and saffron. Goldenrod's mustard-yellow blooms marked unmown borders, swales, and odd corners. A few asters already waved intensely purple flowers.

I was learning to savor September, a subtle month when the sycamores and black locusts dropped leaves. Yellow petals fluttered across country lanes; brown fronds tumbled after cars, rasped on the blacktop, and curled in the parched ditches. The roadside trees were dusty, their heavy foliage sagged, but pastures had taken on a sheen as the worst of the heat lifted. Rains were still sparse but moisture lingered, and the grasses used every drop. Summer had spent itself and fall hadn't arrived. September hung in the balance, a mellow retrospective on the struggle that was history, on another growing season's savage extremes. *What a dry year.* Or, *Sure got flooded.*

I always had to remind myself that September is late summer, not early fall. I'd decided there really were only two seasons for farmers, winter and summer. Spring and fall were transitions, and transitions were killers. Farmers had to reposition, quit what had been working, start over.

I sped down our driveway, looked left toward the Confederate flag marking Johnny's trailer up the road, and then turned right, south. I took the quick jog left at Sam's house and rattled down Snowden Road, trailing white dust. The former cornfield looked good. In August, in pointed disrespect to Fred, I'd hired a farmer who lived on Ridge Road to disk and plant it; soft-leaved orchardgrass, paler green than the region's tough Kentucky 31 tall fescue, now clothed the field. The dark backdrop of woods at the entrance to Mossy Dell had faded from summer's deep green to a more muted shade, faintly yellowish, with scattered russet and orange highlights: the tips of some sugar maples were changing color. The sky above the trees, hazy at the horizon, was a tender blue, layered with white clouds. Frost could come any time, but wouldn't tonight.

I parked in front of the pond and walked uphill into the north pasture, feeling Mossy Dell's hush. At the corral Pete and I built in front of Massey's shed, I watched the sheep graze. Yesterday I'd given them a fresh paddock of Electronet, so they ignored me and ate, filling their rumens before night-fall. Flower crept up for her daily strokes from me on her big, flat head; she stood still, savoring the attention, her amber eyes turning gravely upward to meet mine. I lingered instead of rushing back to the hilltop. It felt like the first time I'd paused since our move to Athens. And somehow, while I was distracted, we'd raised another crop of lambs.

Surely I was a grass farmer at last, which by then I knew meant someone moving forever into unknown territory. Behind this farm, my love at first sight, now so peaceful, was all my work, all my suffering, all my family's suffering. I knew that, at best, things were going well in that moment. The animals were fed and in position to head to the other side of Mossy Dell where there was plenty more grass. This was a day's satisfaction, a sigh. Yet I felt grateful. And somehow cleansed.

Kathy walked up behind me. She'd driven over to check on me, and there I was, leaning on the corral's top board, gazing at the sheep.

"Hi," she said.

"Hey."

She looked at me quizzically as I turned back to the flock. I glanced again at her, a silly grin on my face, willing her to stare at the livestock with me.

"I'm making us an early dinner," she said. "I picked up sausage and sauce for pizza. I found that red wine we like."

I smiled at her. I couldn't speak, managed something like a smile.

Plants and animals want to grow; no miracle had occurred. Or maybe an everyday one had. Watching the sheep, I felt such deep pleasure. Somehow I refrained from pointing stupidly and saying, "Look!"

Kathy saw she couldn't budge me. "I'll go start dinner," she said. "Come home when you're done."

I nodded, lost, staring at the flock, the late afternoon tipping toward evening. Kathy laughed and left me there.

In our third winter on the hilltop, cozy now in our house, Tom and I began getting ready for our annual ritual, Pinewood Derby. In our previous two races we'd grown competitive, living down our disastrous first race when we'd appeared without a weighted car. Last year we had taken seventh place. By now it was a family tradition that each year Claire would scoff at our efforts, and now, in middle school, she was brutal: "It's a bunch of wimp dads trying to be alpha males."

But I didn't care what anyone said about the ridiculous competition. Tom wanted a trophy, and I wanted to help him win one. In January 2001, as our final shot at Pinewood Derby glory approached, with Tom soon to enter middle school himself and leave his childhood officially behind, I pulled out all the stops. I pestered every guy I knew for tips, including the lackadaisical handyman who was still working on our house ("Get a better vise to hold the car," he said) and our perfectionist dentist, who raced go-karts. Although himself the father of two ineligible girls, our dentist had once helped a friend's boy win the Pinewood championship. One day his receptionist called me to pick up Doc's seven handwritten pages of instructions. And I ordered a Pinewood strategy booklet from the Internet.

I tuned our car by sanding its plastic wheels and the shiny silver nails that functioned as its axles; I mounted one front wheel high to lessen friction: two tires were unnecessary in front, my Internet manual said. I filled the precut slot for the rear axle with putty and drilled new holes to lengthen its wheelbase. Using a chisel, I carried out Tom's orders to recess our lead weights into the car's belly. Tom painted it electric blue with white splotches, and named it Clouds Like Blue Pancakes. To nudge its weight up to the legal limit, I used digital scales in Kathy's office and in the lobby of the Athens post office.

By the end, I had spent a small fortune and bore a white crescent scar on my right index finger from an errant stroke of a coping saw's blade. Tom assessed our odds of winning at one in fifty, but wouldn't explain his thinking to me because of my unmathematical mind. I saw things more simply: we were up against two classes, geeks and rednecks. The Californian who'd written my arcane guide hinted at advanced techniques I couldn't perform. I figured we were a long way from Silicon Valley and had most to fear from locals, who wouldn't read such esoterica but who weren't at all embarrassed about wanting to win Pinewood.

I imagined that the husky Scouts who marched to weigh-in carrying their cars inside customized tackle boxes, trailed by cheering sections of youthful grannies and sprung-bellied uncles, were scions of scouting dynasties—big clans with multiple entries every year and basements devoted to Pinewood perfection. Their fathers, the sons of coal miners, used every lathe and drill press and saber saw in their and their buddies' workshops. I envied their roots and sureness of purpose. But I wanted to whip them and get Tom a trophy. This Pinewood would pit my awkward outsider striving against their ability and culture.

In February we won first place at Tom's school, against his pack; his car flew undefeated to easy victories. Though Tom won a fishing pole, the event was really just a chance for everyone to troubleshoot before the official competition, all nine packs in a day-long race-off at the mall for regional honors.

That Saturday morning, Tom won easily again at his pack level, making his car double pack-champion undefeated, a record in our Cubmaster's memory. Tom received a heavy bronze medallion on a sash. That was nice, but we couldn't relax as we ate our burgers in a mall restaurant. Next, the top three cars from each pack would face off in a double elimination.

As our last races began, we worried about a wafer-thin black car and a curvy green one, but they each lost. The competition was so close that Tom had to race a gold-colored rival repeatedly. "The line judge is dumb as a turnip!" I fumed. But after three races, the referee finally called the contest in Tom's favor. Now it all came down to a red car with weights mounted aggressively on its hood, but it had lost once. "Tom will take first place if his car wins this race," the announcer intoned. "If he loses, they must race again. He is undefeated, so he would have to lose twice."

Tom drew the track that had seemed fastest. He lined up his car and tugged its wheels away from its body to lessen friction. The other boy, younger, in his first Pinewood, slapped down his car. When an official dropped the peg and released the cars, they slid together downhill, gaining speed. They were even! And then, near the end, the red car shot ahead. Clouds Like Blue Pancakes had lost, for the first time.

I knew by then what the other car's apparent acceleration meant. Tom's car had slowed, but the other hadn't, or not by as much. A clear and surprising loss. The boys drew lots and Tom got to pick the track. He looked at me, and I shrugged. He thought hard and picked the other side. But the second race was identical: the red car pulled away in the homestretch. We'd been racing for seven hours in the mall's watery gloom, and it was over.

I walked over to the winning boy's parents—actually his grandparents, I thought; they appeared too old to have a child that age. The taciturn man—full mustache, dark flannel shirt, and pale blue jeans—shook my hand. *What did you do?* my pleading face asked. *We can't race anymore—tell us!* Pinewood dads were cagey, or liars, but usually revealed something in victory. Yet he regarded me with a merciless poker face, giving nothing away, not with three years of Pinewood before him.

As my family walked into the mild late-winter afternoon's rinsing breeze, Kathy repeated what Gramps had told her: "I just put the car together. Everybody tried to give me advice. But I didn't do anything."

"Right," I said. "He didn't do *anything!*"

Given all my efforts, what a maddening lie. Or what did he consider "anything"?

"You know he did *something*," Doc said a week later, pausing as he worked on my teeth. He grinned at me like a maniac, his blue eyes bugging as if still enlarged by the magnifying glasses he'd shoved atop his bald head. At the mall, he'd emerged from the crowd to give me a pat on the back and a consoling shake of his head.

"I bet he trued the wheels," I said. "Maybe he could make them perfectly round."

"He did *something*. He had to. You *know* he did." He chuckled and lowered his eyewear.

But I'd never know his trick, never know how we lost, and it irked me. Not knowing stung worse than defeat. Then I got amused. Gramps was even more competitive than I. And his pitiless secrecy preserved a mystery that would make our battle memorable. We'd never forget it; we'd cherish it. I remembered how the man's silent wife had stared all day, as if she'd borne witness to his triumphs many times and was just waiting for another.

They'd competed alone: just the boy, the man, and his ineffable talent. However he'd won, through a poor boy's evolved skill or a master machinist's expertise, or both, he'd earned it. And Tom owned our respectable consolation, now adorning his bedroom shelf beside Clouds Like Blue Pancakes, a damn fine second-place trophy.

CHAPTER THIRTEEN

With my fears, how can I plant all the trees that need planting?
—Farm Diary

A MONTH BEFORE OUR THIRD LAMBING AT MOSSY DELL, GLEN Fletcher called and said he'd decided to move to Montana, buy a piece of land, and build a cabin. "Maybe I'll get a job on a ranch," he said. "Fish in the summer, hunt elk in the fall. Dodge grizzly bears till the snow flies."

"Really?" I was shocked—Glen was such a farmer, such a shepherd—and I felt, at the same time, a twinge of envy. Hunting and gathering and helping someone with chores seemed so easy and peaceful compared with running a farm.

"Yeah," Glen said. "Hank and Mary's kids decided to sell the flock. Are you interested in buying some ewes?"

Is a junkie interested in heroin? It was a chance to get proven mature ewes, and especially tempting given the strange problems with Jo's blood-line. The answer was yes.

"I can't believe you're going to become a mountain man," I added. "I won't recognize you with your beard down your chest."

"Maybe not."

"*Montana?*"

"Montana."

The Goss flock, which had lambed in late winter in the barn, consisted of about eighty-five mature ewes and sixty yearlings. I put 1,000 miles on my truck as I drove back and forth to Dayton, sorting through the flock with Glen's help. "Just keep me fed," I'd told Kathy, and she had.

In the end, I agreed to pay Fay and Curtis Goss $7,000 for ninety-five females and the ram Glen had gotten in Kansas. I wrote them a check for $3,500 up front in late March, with the balance due July 1. Glen helped me find a livestock hauler, who brought the ewes to our hilltop. To cover the purchase, I sold my first sheep customer, Bryce Chatsworth, fifty-three of the Goss ewes on April 1, earning back $5,305—a hefty markup over my own bulk purchase price, but about $1,000 below the going rate for commercial ewes. The sale brought down my total price for forty-two ewes and the ram to only $1,695. By hanging in there and playing my cards right, and with Glen's help, I'd gotten an incredible bargain.

But with the new flock on the hilltop, I felt anxious, and one day noticed I'd chewed my fingernails to stumps again. I'd more than doubled my flock's size, a huge leap forward in my farming career. And I'd brought the hilltop farm at last into play.

Lambing had just begun that April of 2001, and Red was in trouble. In labor for more than twenty-four hours, she struggled and suffered on the floor of Massey's shed. I decided to take her to our veterinary practice for a cesarean section. For any mammal, gestation's dramatic final act is especially risky. I remembered how a friend's wife in Bloomington had almost died from toxemia in late pregnancy—and like that emergency, Red's crisis felt so frustratingly wrong.

There seemed to be a genetic defect in all the ewes I'd bought from Jo. During our first lambing, there was Runt's vaginal prolapse and dead lamb, the ewe that needed a dead lamb pulled, and the pregnant ewe that suddenly died—probably from dietary toxemia, but her underlying issue also might have arisen from a genetic flaw. Last year it had been Cream's strange rectal prolapse. Too many problems from among only fourteen original ewes.

At least my other favorite ewe, The Good Mother, who'd brilliantly raised twins her first and second lambings, had just twinned again. And Freckles, the chunky brockle-faced ewe raised from the batch of lambs Glen had let Kathy pick, also had dropped her second set of twins. But Red's plight consumed me; she needed professional help if she was to survive.

"If you decide to take Red to the vet, don't try to load her by yourself," Kathy had said as she left for work that morning. "Ask Sam to help you." I'd nodded, but I didn't want Sam pestering me and had figured out how to load Red alone.

I dragged our big two-wheeled garden cart from my truck in Mossy Dell's farmyard to Massey's shed. Wheeling it as close to Red as possible, I squatted and lifted her front legs inside. But when I tried to get my hands and forearms under her butt, she collapsed on herself like a sack of potatoes, dead weight. Using my legs, I heaved again, propping her front legs in the cart and quickly lifting and pushing her body inside. At last Red settled heavily into the plywood cart, filling it, and I towed her down the hill to my truck.

Getting her into the pickup's bed involved another lift, also hard in stages. I shifted her limp body quickly upward in a short heave, slammed the tailgate, and headed for town.

The telephone rang shortly after I got home. A girl from the vet's office said to come get my sheep. The surgery was successful, she said, but they needed the space. I arrived to find two large lambs crying for milk in a chain-link kennel. Red was alive but lying unconscious under bright lights on a shiny steel table. Her condition surprised me, but the veterinarians seemed to be having a busy Monday and had no time to talk. Someone helped me carry Red to the back of my truck, and I drove home with her lambs baaing behind me.

Back home I looked in the truck bed at the lambs, shiny black ewes with white spots, and at Red, who breathed rapidly and shallowly, her eyes still closed. We'd have to raise the lambs on bottles, mix up the costly powdered milk replacer, or graft them to another ewe. I couldn't reach Mike Guthrie, and called another experienced shepherd from the grazing group for advice. I told him that Red appeared to be dying. "Don't you know that C-section ewes always die?" Phil said. Another shepherd later told me that he had several ewes walking around with C-section scars.

But Red died. She was dead when I returned to my truck in ten minutes. There wasn't time to absorb the loss. I had two bum lambs—orphans—the season's first bottle babies. I put them in a stall in the barn for the time being. I had to do something with Red's body. As I drove to Mossy Dell, Sam waved me down. It seemed impossible that I'd been able to get past his place four times already without encountering him. There he was, in his creased work pants and oiled leather work boots, a wide smile across his pink face, his blue eyes merry. Sam was always cheerful.

"How's Richard on this fine morning?" he asked. "Does he have lots of lamby-kins?"

"Not yet," I said. "Just getting started." I'd never reveal that Red was lying dead an arm's length from us in the back of my truck. I didn't want to talk about it with him or anyone, except Kathy. She'd understand and share the loss. She'd take over with Red's lambs when she got home, my partner in everything.

In the farmyard, I backed up to a mound of wood chips, dumped there by a tree-trimming company, and rolled Red's body onto the pile. I used a pitchfork to cover her in a blanket of mulch. She'd quickly become compost. It was a mild April morning, suffused with birdsong.

The pain arrived two nights later. When I lay down in bed, an agonizing stabbing beside my right shoulder blade roused me. I almost cried out in surprise. I rose and hunted unsuccessfully for the codeine pills I kept stockpiled for times my back went into spasm. And so I had a shot of 101-proof Wild Turkey bourbon, then another, and returned to bed. The pain forced me back up. It appeared that I couldn't lie flat, but the discomfort wasn't too bad if I was upright. I sat on the couch in the living room and drank more bourbon. As I got drunk, I tested my ability to recline but could lean back only a little. I'd have to sleep propped up. Toward dawn, I passed out. In the morning I'd have a hangover to add to my woes.

Over the next two weeks, taking time off work, I lambed ewes in the daytime, mowed pastures, and drank myself to sleep on the couch at night.

Twice I saw a massage therapist, who pressed hard into the painful spot beside my shoulder blade, trying to force the muscles to release. But I got no relief. I'd never suffered from a pain like this, and never so high in my back.

Two reporters and a camera crew came from an Ohio farm journal and a television show to profile me, a shepherd raising an unusual breed of sheep that didn't need to be sheared. I smiled for the cameras and got through the interviews, but I was getting worried. A month after Red's death, I awoke one Sunday morning with my neck locked and angled forward. My right arm felt like it was on fire, and I had trouble coordinating its movements. In the shower, raising my arm to shampoo my hair, my hand almost missed my head. Pain shot into the muscle of my right chest.

An osteopath injected cortisone into my neck and back, adjusted my spine, and told me I might have a twisted vertebra. But his x-rays showed nothing skeletal amiss. He prescribed Vicodin and Valium for pain and a steroid to promote muscular healing. Finally in late May I saw my regular doctor, who sent me to the hospital for a soft-tissue scan. I would have to lie flat, and still found it terribly painful to do so. In preparation I swallowed a heavy dose of painkillers, and as I climbed into the cylinder I stuffed a handful of hot peppers into my mouth. I'd learned accidentally one day when eating lunch that chewing hot peppers reduced my pain, probably by stimulating my body to produce a surge of endorphins. Between the drugs and my home remedy—and Kathy's sitting beside the tube talking to me about Claire and Tom—I got through the procedure.

The scan revealed the source of my problem: a ruptured disk in my neck was pressing on nerves leading to my shoulder, arm, and chest. Lifting Red, I had blown a disk. I'd protected my lower back, but the spine runs all the way to one's skull. Squatting, I'd loaded my shoulders and upper spine with Red's entire weight. Between the vertebrae, the disks—those cushions in the spinal ladder with the consistency of crabmeat—had compressed. The bases of two vertebrae at the nape of my neck had squeezed down on a disk like two molars crushing a grape. The disk had bulged outward and squirted into my neck. Herniated.

What had I been thinking? I outweighed Red by maybe twenty pounds. I was a weakling compared to the area's burly farmers and stocky feed-mill workers. Men whose upper arms looked as big as my thighs. Men who lifted 50-pound feed sacks with one hand and slung them into the back of pickup trucks. Men who sailed 60-pound bales of hay into barn lofts and wrestled mulish calves and meaty hogs into submission. Men who fought gravity daily with their own thickened bodies and sagging bellies. Strong men.

Men who wore scars from back surgeries.

The next two weeks were a blur as I awaited consultation with a neurosurgeon. I got more physical therapy, and daily life went on, heedless of my pain and dread. I dewormed the flock, moved the sheep to the other side of Mossy Dell, and arranged with a farmer to take the first hay cutting on the former cornfield. Kathy and Claire were raising one of Red's lambs on a bottle, and I'd grafted the other to a ewe already nursing her own twins. The kids were finishing their school year, Kathy was always busy as a dean, and at work I was marketing two new farming books.

RFD was selling poorly, and *The Heritage* modestly, but the press's senior editor and I had convinced our director to publish a revised edition of *The Sheep Book: A Handbook for the Modern Shepherd*, whose author, Ron Parker, I'd met through the Sheep-L forum. *The Sheep Book* was an unusual offering for a small university press, but I was convinced there was a niche for it because competing books were dated. (I was right: it became one of our perennial bestsellers.) And I was pushing another pet project, like *RFD* a bouquet to Ohio's Appalachian region, *The Man Who Created Paradise*. The story, by prolific farm-and-garden writer Gene Logsdon, is a fable about a man in southern Ohio who restores the beauty and fertility of strip-mined farms; the man's efforts stimulate an economic and cultural flowering that heals the region's threadbare communities. Although its text is short, *The Man Who Created Paradise* features gorgeous landscape photographs and a foreword by legendary agrarian writer Wendell Berry.

"Maybe we continue to need to think of Paradise, and of making Paradise," Berry begins, "because the earth as it was given to us (as we realize from time to time) was so nearly paradisal, and we are so talented at making a Hell of it."

Walking across Ohio University's lovely campus, passing through a brick courtyard, crabapple-shaded, to my snug office in Scott Quadrangle, I saw how blessed was daily life. *Any* day without physical pain was good. Get up, have breakfast, go to work; drink coffee and do your job; try a new lunch place with coworkers, gossip and laugh; return to your loving wife, hug the kids, make dinner, read a book. Why wasn't that enough for me, for anyone? Why hadn't I seen how good ordinary life is? Why didn't I ever see life's essential sweetness until something went bad?

I couldn't admit the depth of my fear even to Kathy. And my injury wasn't something I shared with coworkers, though I wanted to yell, "Everything's changed! I'm hurt!" My e-mails to Bailey became cryptic. His reservations about my farming adventure now seemed prescient. Maybe I'd live with pain and impairment. I saw how much I'd risked, maybe how much I'd forever lost. What had I done to myself? Why had I done this to myself? I'd convinced myself I could lift Red, that I didn't need Sam, but now the act just seemed irrational and self-destructive.

Bewildered, I asked a brother-in-law, John Wylie—a Washington, D.C., psychiatrist—"Did I hurt myself to be like my father?" We were talking on the phone, planning my visit to see a friend of his, a spine expert, while I awaited my appointment to see an Ohio surgeon.

"That's bullshit," John shot back. No one would do this to himself consciously, surely, but intention wasn't the issue, as a more rigid Freudian might have told me. Anyway, I *had* enacted my version of my father's dream. Did my injury stem from his genetic weakness, his bad back, or did it somehow reflect my inheritance of his emotional baggage? I recall wondering, really for the first time, whether this was even the dream for me. My self-doubt bordered on self-disgust.

At least the pain in my shoulder and arm was lessening. This seemed great—*I'm healing!*—until I saw John's doctor friend in Maryland at the end of May, and he said the cessation simply meant nerves were dying. Muscles are kept alive by nerves, fed by them, just as roots carry nourishment to plants. When sustained pain kills a nerve, the muscle it supports begins to die.

Two weeks later I saw my Ohio neurosurgeon, a willowy blond woman.

She told me to squeeze her fingers with both of my hands simultaneously. When I did, she raised her eyebrows in alarm. "This is an emergency," she said. She turned to her assistant, who stood by with a clipboard. "Schedule him. Move him ahead of noncritical patients."

I was losing my right arm. My left grip was strong, but my right exerted almost no pressure. Although it had happened so gradually I'd hardly noticed, my dominant hand was weak as a kitten's paw. My right arm was measurably smaller than the left, too, withering as muscles shrank. In the doorway of the assistant's office, I stood slump-shouldered beside Kathy. The nurse consulted schedules, looked up from her desk, and said, "She can operate on you in two weeks."

Bailey called a few nights later. "I couldn't tell from your e-mail whether this is routine surgery, as if there is such a thing, or if it's the big banana."

"It's big," I said. "I've been getting hurt a lot, but this is different. Bailey, you were right."

"You're being hard on yourself. I think we both like being overwhelmed. It validates us."

"But the farm still isn't making money. There's no reason to do something this hard."

Alarmed, he e-mailed Kathy: "Richard sounds depressed. At least he finally seems to realize he can't go on like this. I've got too much on my plate, too, but I don't have a hundred sheep and a bad back."

Kathy forwarded me his note, but didn't want to discuss it when I brought it up the next morning before we left for work. "We just have to get through this, Richard," she said, shouldering her bulging briefcase in the kitchen. "Getting depressed isn't going to help a thing."

"Thanks for the sympathy," I spat. She started to speak but just shook her head, and left.

She didn't understand my constant pain and fear—and hadn't tried to! Wouldn't a good wife try to comfort her husband? Stony with anger, I drove the kids to school. When Kathy called later that morning to ask if I wanted to get lunch, I told her I was too busy and almost hung up on her. Two hours later, she appeared in my office's doorway. "Come on," she said. "I'm buying."

I hunched in the red vinyl booth in the town's Chinese restaurant. The waitress set a plate in front of me and gave Kathy a steaming bowl of egg drop soup. "Talk to me," Kathy pleaded. It would've been a good time to try, but I couldn't speak. Anger constricted my throat and caused it to ache. I stabbed my Kung Pao chicken. We didn't fight often, but when we did, this sometimes happened. Kathy would recover quickly and move on, but I'd get so angry that the emotion would take over. I'd spiral into a foul mood that might last for days. I hated this about myself—and feared it, the anger I carried—but felt helpless when in its grip. I stared at my plate, pushed food around with my fork, choked down some meat. Kathy sipped hot tea and looked out the window.

By the time I picked up the kids at school I had a plan. I drove them home, changed for chores, and put my chainsaw on the passenger seat of the truck. As I turned down Snowden Road, I saw Sam Norton running across his lawn to catch me, but I stared straight ahead and gunned my pickup past his house.

After I'd fed Flower and given the flock an extra-large break of grass, I stood in Mossy Dell's farmyard, looking at the tulip poplars behind the cabin's old site. One of the trees leaned away, trying to catch sunlight. That would be the one to drop, because I could be almost certain in which direction it would fall. And I could be standing there, right under it. It would appear to be an accident: just another farmer who got caught while felling a tree. Happens all the time. I'd be a suicide, like my grandfather. But my troubles would be over. Kathy could move to town, where Claire and Tom would be better off, close to school and friends.

I opened the truck's door and gripped the orange chainsaw in my good hand. It would be hell to use. Suddenly appalled and frightened by my self-pitying melodrama, I released it.

That evening at dinner, having shocked myself almost into normalcy, I spoke to Kathy—routine stuff. Not about my despair. I felt emotionally hungover, chastened.

Suicide had so scarred my father that it shamed me that I'd pretend to consider it. Dad was fourteen years old on the Thanksgiving morning when he went looking for his father, Charles, in the silent Detroit mansion. Home from boarding school, he'd risen early because Charles had promised to take him rabbit hunting at their farm. He pushed open his father's bedroom door, left ajar, and saw the empty bed. He found Charles crumpled on the bathroom floor, a shotgun beside him on the bloody tile, his head shattered from the blast.

That weekend Dad was hunting alone at the farm, out in the woods, when his back went into spasms. He had to crawl back to the farmhouse in the snow.

"Your father's life," Mom told me when I was a boy, "is the saddest story you'll ever hear."

At age four he'd been in a train wreck with his mother, Eva, that had disfigured her face and severed one ear; doctors fashioned a flap of flesh as replacement. Charles, abandoning Eva to her pain and grief, moved out of her bedroom. "You aren't beautiful anymore," he said, according to Dad's older sister. Over the next decades Eva underwent thirty-seven operations on her face. Dad took a nightly loyalty test as a boy when his parents separated after dinner. Which to follow?

After the stock market crash of October 1929, Charles spiraled into depression and curled like a baby on his bed. He was treated by a therapist whom Eva later investigated and exposed as a charlatan. The man wasn't a Vienna-trained psychiatrist, a disciple of Freud—he hadn't even been educated as a doctor.

The day of Charles's death, Eva told the *Detroit News* that he'd killed himself because of poor health. But that story failed to lessen the stigma for his family; suicide was a blow that compromised their social status. Dad returned to boarding school, where I imagine he cried alone in the night; maybe that was where he learned to get along without other people. He never spoke of his father—but once, as we watched television together, he said out of the blue, his eyes blazing, his face an agonized mask, "I hated him for years." As my sister said to me when I was an angry teenager and criticized him, "If you crawled inside his head you couldn't stand the pain."

He didn't respond to life like other people, so everything he did and said seemed significant to me, and I monitored his activities. The place in our garage in Satellite Beach where he polished his shoes was sacrosanct and fascinating. Other fathers talked and did things with their families, but not my father. Other families took vacations together, I noticed. I envied their togetherness but knew that was not his way.

When I was a boy in Georgia, Mom once ordered Dad to punish me for pulling all their books from the shelves. She was pregnant and didn't feel like whipping me herself. He spanked me, went in their bedroom, sat down, and cried. "I've proved I can beat up a five-year-old," he said. She didn't ask him to discipline me again for years. He would have been a nice balance to her hot temper, but he was a rare visitor to the domestic sphere—in a relationship with his Rosie, but relating to his children only through her.

As a kid I imitated the private father, the real one. The one who was sad. Whenever I provoked Mom—which was a lot—by sassing her or sulking or tormenting my siblings, she'd whip me with switches or belts or hit me with her bony hands, and once she thrashed me in the garage with a handy garden hose. I was shy and sullen, plagued by nightmares and nosebleeds. I wet the bed. In school I daydreamed and doodled on my assignments. Eventually, I concerned Dad, and one morning over breakfast he suggested I read our newspaper's comics page.

When I was a teenager, one Sunday afternoon he protested, as if reading my mind, "I'm not an unhappy man." I wondered, did two negatives make a positive? I wasn't buying it. He couldn't soften my inescapable conclusion that we'd lost something huge, Stage Road Ranch, and were toughing it out, like the broken characters in the Hemingway novels I was reading. When he seemed happy, he was only betraying our truth, pretending for outsiders, puzzling his other children, and infuriating me with his hypocrisy.

His leisure activities with his kids—taking us water skiing and fishing—stopped when I was twelve. He quit after his first heart attack, which almost killed him on Thanksgiving Day, 1967, the thirty-fifth anniversary of his father's suicide.

One afternoon before my surgery, I noticed that a Goss ewe on the hilltop had a swelling behind a front leg. The lump dripped greenish pus. The growth was the size of a lemon, and the skin across its front wall had thinned and burst. I checked the flock and saw two more ewes with similar lumps, not yet broken.

I called Glen, who was spending his last summer in Dayton, building barns, and he said "boils" were a regular occurrence in the flock. He used to lance them with his pocketknife. He was matter-of-fact about the problem: just something sheep had, another shepherd's task. But I realized this was a disease, one I'd read about. It carried a formidable name: *caseous lymphadenitis.* The words describe the problem in the literal, precise, and poetic nomenclature of science. *Caseous* means cheese-like, which describes the paste of dead tissue inside the lesion. *Lymphadenitis* means an inflammation of the lymph nodes. The disease is abbreviated to CL by shepherds, or called cheesy gland, or boils. In addition to being disgusting, it's highly contagious. One ruptured abscess releases billions of bacteria. And the germs can live for at least five months, maybe as long as a year, in manure, bedding, corrals, and soil.

In CL's internal form, abscesses can infect the liver, lungs, spleen, kidneys, and spinal canal. Glen had sometimes lost sheep in their prime, and now I knew why they wasted to skin and bones. He'd shaken his head once when showing me an emaciated ewe penned for shipment to the livestock auction to become pepperoni, the fate of spent ewes. "She's got parasites," he complained, "and I can't knock them out with this drench." It was now clear that he'd been fighting the wasting disease. Slowly deadly, it preserves its unfortunate hosts long enough to spread itself to lambs, other ewes, and rams.

There was no way Glen could defeat such an awful disease—not with his worm drench. Not with his knife blade as he pinned a ewe in a cobwebby corner with his shoulder. And this explained the little yearling I'd lost last summer, one of the ewes we'd bought as a lamb from Glen. She'd aborted in the winter, two dead fetuses in the pasture. Sometimes abortions just happened, I'd read—a normal occurrence. But then, instead of fattening without lambs to nurse, she lost weight all spring and by June was so thin I could see her ribs. I'd worried about her, figuring this was my own recent parasite problem, and had given her extra drench. But she was staggering by the end of the month and dead by the Fourth of July. Busy moving the flock across Mossy Dell, I left her body where it lay near Massey's shed. There was hardly enough left of her to keep the buzzards busy for two days.

Now I researched CL in my books and online. The surface form gains access through the skin, entering minor cuts, or through abrasions inflicted as a sheep vigorously scratches an itch on a manger or fencepost; the internal ailment is spread when a diseased sheep coughs a mist of bacteria into another's face. In a flock that's closely confined to barns for long periods, as the Goss flock was during the winter and in lambing season, conditions are ideal for transmission.

Australia, which takes sheep seriously, had made an effort to eradicate cheesy gland. The manufacturer of a vaccine claimed that CL was once present in 97 percent of the nation's sheep flocks; losses from condemned meat ran each year to almost $20 million. In the United States, at least half of flocks in the East were believed to be infected. I had to assume that CL was entrenched, given the flock's close confinement history and Glen's well-meaning but ineffective treatments.

I reflected on something else: cheesy gland can spread from animals to humans. Such infections are rare, I read, but have been documented outside the United States where people have gotten cut while skinning infected sheep or goats. A few people might have caught cheesy gland from drinking raw milk, presumably where internal abscesses were draining into an animal's udder.

Though injured myself, I had to act. I caught the ewes with lumps and isolated them in the barn, mindful that I'd have to diligently clean out the bedding. Our older Goss ewes were already effectively in quarantine on the hilltop, as our original flock and rams were at Mossy Dell. That separation, a lucky accident, was alone on the positive side of the ledger. Mossy Dell's flock appeared clean, and there was relatively little contamination there from the one casualty, since she'd had the internal form.

My liabilities were easy to tally: a herniated disc in my neck; a rare genetic birthing problem in my original ewes; and in my new flock, a nightmarish disease.

My surgery was scheduled for Friday, June 29, the day I'd marked on my calendar to mail a $3,500 check to complete my purchase of the new ewes.

The surgeon cut a slit about two inches long in the front of my throat, slightly to the left of the midpoint, in the seam where the neck meets the chest, avoiding the voice box. I'd signed releases acknowledging various risks, including the loss of my speaking ability and even death. She worked her elegant fingers past my esophagus and arteries and removed the damaged disk. In its rightful place between two cervical vertebrae she inserted a wafer of thigh bone harvested from a cadaver, which would become part of my spine as it fused the vertebrae. The process reminded me of how nurserymen graft a branch from one apple tree to another. Often, instead of a bone graft, surgeons screw titanium braces to the vertebrae, but I'd opted for the more solid but slower healing repair advised by the doctor in Maryland.

My choice meant I'd live in a neck brace that summer. For six weeks, the brace would restrain my head and neck while bones fused. For six weeks, Kathy would have to care for me, the children, and the sheep while holding down her demanding day job. Before surgery, I'd strung as much electric fencing as I owned, staking out successive paddocks where Kathy could move the sheep as they ate their way through the pastures. There was little I'd be able to do after surgery, with the brace restricting normal movement. And anything jarring was out of the question. No walking into pastures at the risk of stepping into a hole. No driving.

The surgery saved my right arm, though it was impaired and I'd have to strengthen it through weight lifting. I'd lost too much time with therapists and homeopaths who'd treated symptoms. My triceps muscle was almost gone, withered by the loss of the nerve that carried the pain; a divot of muscle was missing from the pectoral of my right chest.

Kathy brought me home to the hilltop, and I took two Vicodin and rolled into bed. Getting in and out of bed is a lingering memory of that blurry time, because I had to learn to move without leading with my head. With any movement, especially rising, I felt as if my head might topple right off my body. The brace produced stiff, mechanical movements, and I could see Claire's and Tom's concern in their eyes. Just making the trip from our bed to the dining-room table felt like an immense undertaking.

Somehow I got through those long summer days. At first, I slept. Yet my farm calendar for that July notes that we received rain two days after my surgery. Three days later, we got another rainfall, and I sent Kathy to the rain gauge to tell me the exact amount, which I recorded.

On some afternoons Kathy would return from her job, do chores, and announce that we were going for a drive. Even my riding in cars was forbidden, but she took me into the country to perk me up. It helped, seeing my favorite season spool past beyond the blacktop, hot and sunny, the grass mown and daylilies in bloom. I clung to the car like a frog gripping a reed in a storm. If it appeared we were going to hit the slightest bump, I stiffened my legs against the floor. Kathy endured my panic. Busy holding together our lives and the farm, she was trying to give me hope and a sense of normalcy. She later told me she didn't know how she got through that period.

My invalid status gave me plenty of leisure to worry about the disease in my new ewes. And the defect that appeared to be fixed in the ewes I'd bought from Jo. A bad genetic roll of the dice, maybe. More likely she'd bred the same ram to his daughters, granddaughters, even great-granddaughters, concentrating flaws through inbreeding. Serious breeders avoid that and track animals' ancestry, their pedigrees. But most people aren't serious breeders. Most are propagators, merely increasing a breed's numbers.

And what of my own genetic flaw, which my foolish act had brought to the forefront, a weakness I'd denied but which steadily intervened between me and my farming dream? I knew I had a bad back, but in one morning I'd thrown aside precaution. The long-waisted Gilbert men—poor conformation!—also suffered from degenerative disc disease: those crabmeat spinal cushions desiccate into beef jerky, and vertebrae grind against each other. In the wake of my injury, x-rays and scans of my spine showed substantial deterioration: bone spurs, arthritis, bulging disks, narrowing of the channel for the spinal cord. Dad had kept a complex traction device affixed to his bed, and once I'd seen him lying on our family room floor, curled in a fetal position from pain, waiting for an ambulance. Mom thought his spine fused naturally in his old age, the vertebral links becoming a single unit, like one of Massey's rusted logging chains I'd found in the dirt at Mossy Dell. At last, Dad's entire back rigid, his right leg withered, he was free of pain.

Unable to root out the bad genetic information that underlay my injury, I vowed to do something about the problem among our sheep. In my own case, I couldn't help but think that if I were a sheep, an elite breeder's verdict would be swift. I'd have to be culled.

I unfastened the Velcro clasps of my neck brace on August 10, 2001, and immediately I put it back on. I'd grown dependent and had to wean myself from its reassuring grip. A week later, the surgeon x-rayed my neck, and after a moment of panic when she looked at the image—she'd forgotten she'd used a bone to splice vertebrae and had expected to see steel—pronounced the graft successful.

"Why doesn't my neck turn?" I asked. "It's stiff."

"That's just muscles," she said, and whipped out a prescription pad for a new round of physical therapy.

A month later, the evening of the terrorist attacks in New York, I went to bed feeling sick and had strange dreams. I'd awaken, then slumber and dream again, unsure whether I was awake or asleep. I seemed to blend waking thoughts into dreams in a shadowy collaboration between realms. In one dream an elderly professor I barely knew appeared, but his presence was the message, and I knew what it meant. Sober, loyal to the university, a fine teacher and mentor to students, he was a good citizen of Athens, of the nation, and of the world. His integrity and concern for others were honored by anyone who had any sense. He seemed committed to things larger than his own desires, to community. I felt I'd been the exact opposite in my fevered pursuit of my dream. I had to change.

We'd been through a lot in our relocation, but now this place was home. I'd shed blood. And the kids would be with us for only a few more years. Claire and Tom were thriving again, and Kathy believed that they were doing better academically than they would have in Bloomington's rich school system. Here they were being treated as special, in programs for academically advanced children. This couldn't lessen my feeling that I'd neglected them. I pictured the raspberries I'd planted for Claire at Mossy Dell, and my asparagus patch, which I'd been too busy to tend and had let fescue and drought overwhelm. And my nursery over there was weedy beyond belief.

I'd failed on several fronts, and for the same reason that most farmers are unable to follow textbook procedures. Already too busy, I'd enlarged the flock too fast, ahead of my skills and infrastructure. I'd been distracted, stressed, nagged by injuries. In the wake of such trials, I might have felt battle-tested, a real farmer; instead, I was deeply discouraged. For all my frustration with incompetent and ignorant farmers like Fred and Jo, clearly I'd joined their ranks.

With my passion for plants and animals sorely tested on the commercial scale, farming seemed unreasonably difficult, an expensive lesson in humility. I resented that something with such intrinsic appeal and necessity was so difficult. "Why," I demanded of Mike Guthrie, "do you have to be a freaking genius to be a successful farmer?" He just shook his head and smiled.

While still laid up, I'd ordered vaccine to fight cheesy gland. I understood that more of the ewes around our house would get sick. The vaccine probably wouldn't help a ewe already infected, but it was supposed to reduce new infections by as much as 80 percent. And I knew that some of the sheep at Mossy Dell might carry CL, because I'd raised them as lambs from the Goss flock, and of course there was the yearling that died.

Even as I planned my campaign, I saw that my reaction to the disease—as awful as it was—seemed excessive. This came from my emotionalism, I knew, and also from my daily dose of Sheep-L's fanatics, who took diseases seriously, in implied contrast to the attitudes of typical shepherds and know-nothing farmers. All the same, one shepherd on the e-mail list, a woman in western Pennsylvania, had cheerfully revealed that her flock carried foot rot, an admission in that company like telling coworkers you had a venereal disease. "I am a commercial shepherd," she wrote. "I have too many sheep to cure foot rot. It would be too expensive and labor-intensive to try. I live with it." This provoked a flurry of snide or condemnatory e-mails. "Do you charge extra for foot rot when you sell people sheep?" a guy in Connecticut asked. She replied, "I don't sell breeding stock. Like I said, I sell meat. Or don't you know what 'commercial' means?"

I e-mailed a prominent shepherd, a woman in Wisconsin, who'd mentioned on Sheep-L that she'd eliminated CL from her flock. "I had to get rid of it when I decided to start selling brood stock," she replied. "It's nasty, but I wiped it out in four or five years by culling and vaccinating. It wasn't hard. I don't know why more shepherds don't fight it. You can do it."

With Sam's help I began vaccinating the ewes on the hilltop and the Mossy Dell flock. We inoculated the Goss ewes twice, and the rams, ewes, and lambs at Mossy Dell four times. To avoid spreading the disease inadvertently, I changed the needle in my vaccination gun with every sheep I stuck in the neck.

At the end of October, Jim backed his long aluminum stock trailer toward Massey's shed. We loaded the sheep, hazing them aboard through the corral Pete and I had built two years before. I had repeatedly vaccinated at Mossy Dell partly because I'd decided to meld our two flocks.

When I'd visited Jim at the barbershop and asked him to transport our sheep across the road, I'd said I was running from parasites, and that was partly true. Mostly I wanted to simplify while I fought Jo's genetic problems and the Goss flock's disease. Doing so, although achievable, would take time and effort. Of course, after one grazing season without sheep, most of the parasite larvae waiting in the valley farm's pastures would be dead. And the former cornfield that enlarged it, though still unfenced, was clean, with acres of sweet orchardgrass for grazing.

My only thought, though, was to retreat. As Jim swung his rig wide into the farmyard to enter Snowden Road, I noticed the bright green foliage of one of my chestnut oaks shining in a shaft of sunlight on the hill where the cabin had stood. Beyond, the old farm's empty pastures to the south had regrown, a month of good feed. But I turned my face toward the hilltop. I rode away from Mossy Dell in Jim's diesel pickup without a plan for when I would return.

PART FOUR

A WAY TO BE

There is no doubt about it, the basic satisfaction in farming
is manure, which always suggests that life can be cyclical and
chemically perfect and aromatic and continuous.

—E. B. White, *One Man's Meat*

CHAPTER FOURTEEN

Old Mama's triplets total almost 28 lbs.!

—Lambing Notebook

ONE AFTERNOON A WEEK BEFORE THANKSGIVING, A MONTH AFTER I'd moved the sheep from Mossy Dell, Mom called me at the office from our house with news to report: "A man was just here asking for you. He wanted to check that you let him hunt, because your neighbor is upset."

"What was he driving?"

"A big green truck."

"He lives on the other side of Lake Snowden. I said he could hunt deer at Mossy Dell."

Our upset neighbor had to be my irascible barber, Ernie. It didn't take much to rile Ernie, but why would he care if I let someone hunt on *our* land? I picked up the kids at Athens Middle School and drove home, worrying about getting my chores done before dark. In November I lost my light early, especially after the time change, which gave me more daylight in the morning when I didn't need it and then cheated me out of a precious hour after work.

After throwing on coveralls over my town clothes and donning rubber boots, I moved the sheep to a fresh paddock, and fed and petted Flower. I gave hay and water to the rams in the barn. Thanksgiving was my goal for turning them in to breed. Before then, I'd have to sort the ewes, and was still deciding on paper who went with which ram. I was still getting the feel of having the entire flock, fifty-two ewes, on the hilltop. We were more than fully stocked for only sixteen acres of pasture: by Mike's rule of thumb our maximum flock should've been forty-eight ewes, three on average per acre. But the barn was full of hay, square bales from our first hay cut from the old cornfield, to supplement stockpiled grass.

When I cut the floodlights inside the barn, the western sky framed in the building's sixteen-foot-tall back doorway was an incandescent expanse of lemon yellow tinged with brass; cloudbanks, bruised blue and lavender, lay above a glowing orange band that spanned the horizon. A cool breeze, refreshing after my work, rose up the hilltop; dry leaves rustled against the barn's tin wall. The darkening sky above me, and as far as I could see to the east, was marbled with low gray clouds, faintly lit beneath by the sunset.

Our van climbed the driveway, swung to a halt across from the house, and Kathy climbed out. She hurried across the gravel, unaware of me in the dark. I heard the front door open, and saw Doty and Jack trot out in a wedge of light. Doty squatted to pee in the middle of the driveway, and Jack streaked down the gravel toward me. He sniffed my boots and ran to the corner of the barn and lifted his leg dramatically, looking up into my face. I gave him the expected praise—"Good boy!"—as if that would strengthen his shaky housetraining.

I loved that farming sent me outside every day in all weather, no matter my mood. Inside the house, it would appear pitch black outside, though I knew it wasn't; it would seem cold, though I knew it was mild. I kicked off my Wellies on the porch and came inside to the smells of Mom's cooking. I hugged her at the stove, where she stood holding a wooden spoon, stirring the lamb shanks she'd prepared in thick brown gravy. Kathy was setting the table.

"You guys should look at the sunset," I said.

"I know," Kathy said. "We've been watching it."

"Okay," Mom said, "everybody come on!"

Claire and Tom emerged from their bedrooms and we gathered at the table. When Mom visited, every night was a feast. At eighty, she still worked like she was sixty. Twice a year she came north, stuffed us with home cook-ing, cleaned the refrigerator, rearranged the kitchen, and filled our freezer.

"There was more excitement here this afternoon," Mom said, lowering a steaming bowl. "Your neighbor Ernie showed up, right after I called you. I thought we were going to have a shootout right in the driveway if the other guy came back. Ernie knocked at the door, raving. He said he was going to 'return fire.' He looked like he was about to have a heart attack. He said 'your' hunters had crossed onto his land and were shooting toward his house."

I pictured the barber's dark eyes, wild and mournful. "I'm sorry you got caught in the middle of that. What did you say?"

"I asked him, 'Do you have heart trouble?' He said, 'How did you know? I've got so many health problems . . .' I told him, 'My husband died of congestive heart failure, and I know the symptoms. Your color doesn't look right.' I got him inside, sat him down, and heard his whole story—and his wife's—and served him caramel pie and coffee. Do you know his wife?"

"Yes. They live just around the curve. Janet is as sweet as Ernie is angry."

"Her health isn't good, either," Mom said. "Maybe we should take her some chicken soup."

Kathy and I laughed; across the table Claire grinned, knowing the play-ers. Tom concentrated on his plate: he loved his grandmother's cooking. Finally, Ernie had met his match. Mom was renowned for getting people's stories, and by giving Ernie attention, by showing him that simple kindness, she'd soothed his riotous soul.

"You made a friend for life," I said. "But I may have to smooth his ruffled feathers. Tomorrow I'll visit the barbershop."

The next day was Friday, my farming day, and after one of Mom's big break-fasts I drove into Athens. The barbershop would be packed by afternoon with college kids getting haircuts before leaving on Thanksgiving break. Now, at ten o'clock, Ernie and Jim each had a student in their chairs; two more waited against the window wall in brown plastic bucket seats. I aimed for the chair across from Ernie, just inside the glass door, and hailed the barbers.

"I met your mother yesterday," Ernie said even before I sat.

"She told me. She said you were upset about Ed McNabb hunting at our farm?"

"I found him and his sons over there, in the corner of your place. They'd killed a buck on my side and dragged him over the fence. There was hair on the wire."

"Ed's been asking me if they could hunt since we got that land. I finally said yes this fall. Since Jim helped me move our sheep, there's no livestock over there."

"I heard a slug go over my house yesterday." He imitated the high whistle. "I've got Janet there, and my dogs. She's not well and doesn't need to be upset. I won't stand for it. That shot came from my own woods!"

"Sorry, Ernie. I'll put an end to it. He has other places to hunt."

"He's one of Fred's buddies, you know. Part of his 'hunting club.' They drive around, drink, shoot up everything."

Ernie swiped his comb across his student's hair. Jim, his back to us, was clipping around his client's ear. Then Ernie said, "Janet and Fred's wife are cousins, you know."

"You're related to Fred by marriage?"

"Janet and Dolores grew up together. But when Fred and Dolores started dating, Janet went her own way. She never liked Fred's big talk."

I was astounded by this news—would I ever know enough history here?

Ernie said his feud with Fred Paine started not long after Fred and Dolores moved into their brick house, now ours. Ernie wanted to fence for his cattle behind Fred's—I imagined that Ernie, irritated by his ostentatious new neighbor, was marking his turf—but Fred refused to pay half. Under Ohio law, neighbors are supposed to contribute equally toward construction, maintenance, and repair on border fences. To force Fred to obey the law, Ernie took him before the township trustees. Fred just faced them down.

Then, when Fred started fencing his own place, he tried to make Ernie share the cost on two line fences. "That's when I hauled those cars over there," Ernie said, "to spite him." Last summer, as a gesture of friendship to me, Ernie had shoved the vehicles into his woods. Now our house's northern view, my favorite, was of his steep, overgrown fields, his rusty grain silo, and the road curving into the distance below.

Then I remembered that Ernie had finally been inside our house. I felt self-conscious about its beauty: the gleaming wood floors, Kathy's antiques, the Persian rugs Mom had given us, Dad's landscape paintings. "I wish we'd known what was underneath the bricks of Fred's house," I said. "We sure got in over our heads."

"Yeah, you really overhauled it," Ernie said. "That's why my taxes went up."

It was sometimes hard to tell where I stood with Ernie—with anyone here.

On a cold morning in late winter I was driving home to the farm after a Friday breakfast date in town with Kathy. Muslim students were coming to kill sheep. This was Islam's highest holy day, the Festival of Sacrifice, and would be a big feast night after a long day of fasting.

Eid-al-Adha commemorates the willingness of Abraham to sacrifice his beloved son to Allah. At the last moment, Allah allows the substitution of a ram. Traditionally, in a symbolic reenactment of Abraham's obedience, a Muslim family slaughters a large animal. The family consumes a third of the meat, gives a third to friends, and shares a third with the poor.

It was February 2002. As I drove home I listened to the news on my truck's radio about the murder of Daniel Pearl, a *Wall Street Journal* reporter who'd been kidnapped in Pakistan. Warm and well fed in my truck, sorrow stabbed at me. I imagined how cold and hungry and terrified he must have been. I hoped he'd prepared himself. Yet how can one wish to live, to be spared like Abraham's son, but be ready to die? For Daniel Pearl, life's hardest spiritual task was compressed into hours.

The radio announcer didn't say how he died but implied it was horrible, gruesome. So not a bullet. Then I realized. *They cut his throat. That's how Muslim extremists would kill a hostage.* Such a death also would have maximum horror for Americans. I wished I couldn't imagine it, but I could—the Muslim students had come last winter and killed three lambs. I'd been shocked by the intimacy and blood of the slaughter. They'd cut our lambs' throats from ear to ear, and the animals' deaths, while quick, weren't instant. I'd spared a big wether they wanted, unwilling to see any more sheep die. "He's already sold," I lied. After I sent the students on their way, I released him from the barn and he bounded for the flock. Now he was a companion for the rams and enjoyed his carefree, sexless life; Claire named him Agador. But even as I'd pardoned him I knew he was an exception. I couldn't keep non-procreative sheep and remain a farmer.

I'd gotten the Muslim students' business by posting lavender fliers around campus for "lean, healthy lamb." Because I tried to send an evenly sized lot to market every fall, I always had some larger and smaller lambs in the barn. And this winter, Cream.

The students' leader, a man named Jamal, had called me again the previous week. "We would like a big sheep this time," he said. "And maybe two or three lambs. Do you have a big sheep?"

"Yes," I said. "I have one."

Cream possessed personality—those sneakily comic yet vicious ambushes of Flower—and she had sentimental value as one of the fifteen ewe lambs Claire and I had hauled from Maryland to start our farm. But the rectal prolapse she'd suffered meant she also had the genetic defect. I couldn't keep her offspring, or sell them for brood stock, so I hadn't bred her. She was on death row in the barn. Yet, having offered her, I dreaded seeing her die.

There was another matter: letting Jamal kill Cream on the farm would make me a lawbreaker, a real scofflaw. The ancient, elemental activity we were again planning was illegal in Ohio. The state forbade on-farm slaughter, for sale of meat, without a formal, expensive, and inspected abattoir. Although the government seldom enforced it—and farmers around Athens ignored it—Ohio's law was among the toughest in the nation. Ostensibly enacted to protect public health, such rules support corporate farms and America's centralized meat-packing industry.

The law also got the killing out of sight. It seemed to me that the farther Americans were removed from the fact that they live by death, the more ignorant and sentimental they'd become. And nothing's more cruel than a sentimental nation; its people can believe that war isn't hell but something that happens, at a great remove, to other people. Not even to humans, really, but to video-game projections of fear and disgust. I lived closer to death than most Americans but on this point was a hypocrite too. So as not to witness the death of lambs for our table, I took them to a rural butcher and returned for packets of meat wrapped in white paper.

All my life I had thought my father was sentimental because he wouldn't eat the cattle he raised when he was ranching. I'd been learning myself that the nurturer's role doesn't stop suddenly or end cleanly. In October, when a livestock hauler had gotten his aluminum trailer loaded at the barn with my market lambs, what a mistake it had been to glance inside before he pulled away. The lambs had peered up at me as if to ask, *What now, Boss?*

My guilty feeling—*I'm Judas, betraying them*—had surprised me, another farming issue I hadn't anticipated. I hadn't foreseen that the end result of tending a flock of ewes and raising their lambs would carry this inescapable weight. Now it was clear. How could anyone remain unmoved by his role in systematic death-dealing?

And death, in one form or another, always draws near a farm with livestock. Glen Fletcher once saw a coyote attacking the Goss ewes and said it was like watching a border collie herd sheep. The hunter approached, head low, eyes scanning the flock. The sheep bunched together for protection. The coyote selected one animal and rushed in to make its kill as Glen ran for his rifle. Sheep make a perfunctory struggle but, fatalistic and ill-equipped, seldom fight back. A good ewe may defend her lambs from a predator, but only to a point. If she can't save them, she'll save herself. The flock settles, without lasting trauma, the moment a kill consummates the event, the scripted climax of an ancient relationship between predator and prey. My role in this drama was odd, dual: protective shepherd and super-predator, supplying food for his species.

As I drove up our driveway I felt nervous. Very soon I'd be surrounded by young Muslim men wielding knives. It embarrassed me, my fear, but only five months ago, Muslim terrorists had attacked America. I tried to disregard my feelings—I had to get through this. I parked and went in to made a pot of coffee, a comforting ritual, something against the cold. My knees were sore and my spine felt achy, and I reminded myself to be careful handling strong animals.

I heard a vehicle rattle up the driveway and stop in front of the house. It was Jamal's boxy blue van. I hurried out, and as he rolled down his window I greeted him, a stout man whose smile broke like a cresting wave through his bushy black beard. Three younger men gazed at me.

At the barn they produced an assortment of cleavers and small knives they'd gotten from the Odd Lots store in Athens. I helped them sharpen their cutlery and loaned Jamal my sheath knife. I thought about mentioning the troubles in the news, then decided not to—and then did. "Has there been any backlash?" I asked. Jamal shrugged and said, "Nothing serious."

"Religious extremists in all countries bring pain," I offered.

Jamal nodded. "People are people," he said. "Life is short. There is enough sadness in the world. Why bring more?"

I had run Cream in with some wethers, and I shooed her into a corner, haltered her, and led her toward the barnyard. She bucked, tugged against the nylon lead rope, and leaped past me. Two of Jamal's helpers grabbed her and helped wrestle her to the edge of the gravel. The men laid on her back to force her down as I held the rope.

Jamal pointed my knife northeast, toward the rust-streaked roof of Ernie's silo that marked the point where Ridge Road swung away from Ernie's weedy hills and disappeared into trees. "That's east," he said, more confirmation than question. Muslims faced animals they slaughtered east, toward the holy city of Mecca, to honor Allah. "Yes," I said, trying to think. I wondered, as on their first visit, where exactly east was: the sun did rise farther north in summer, so was that due east?

Jamal grabbed Cream's jaw and pulled up her head, baring her throat. Uttering words, a low guttural sound, he plunged my knife into the far side of her neck. Cream jerked and blood sprayed into the air. Jamal sawed the knife across her throat, severing her airway and arteries. The blade clacked against bone, her cervical vertebrae. He rose, having almost decapitated her.

Cream gasped from her cut trachea. The rasping horrified me, and I glimpsed her windpipe's pale stump quivering in her throat. Her blood spurted onto clumps of fescue, still green. I felt panicky—*she's still alive, suffering. I'm a murderer*—and Cream's head fell to the ground. She was unconscious and then dead.

The men were a likable bunch and worked hard in the cold, still morning. It was quiet in the farmyard under the winter sky, a milky blue streaked with clouds ripped in long mare's-tails. They hoisted Cream's carcass using a rope I threw over the front support beams of the dilapidated shed nearest the barn, and they skinned and disemboweled her. They rinsed their knives in a bucket of chilly water I hauled from the barn's hydrant.

Afterward I helped them kill three lambs. It wasn't as bad; the killing was becoming routine. In the best sense, that meant professional: humane treatment and competent processing. Yet professionals risk losing something; there's always truth in a novice's gut response to a profession's exigencies. At a farm conference, I'd heard a shepherd announce how he dealt with triplets, which he didn't want: "I have a hammer, and I know how to use it." Was that the price of commercial shepherding, crushing the skulls of surplus newborns? No, something seemed wrong with that man's world, not just with his practical but heartless act. The Muslims had a stricture against killing animals in front of their kind, and they always asked me the age of lambs because they were forbidden to kill nursing young.

What nagged at me was my silence—again I'd stood mute as my lambs died. I felt impoverished beside Jamal, with his rituals and prayers.

All day, waiting for Kathy and the kids to come home, I thought about Cream. I decided that what caused my pain, my love for these sheep, was also the only thing that would make the killing bearable. I could raise them humanely, let them express their nature in green pastures, even as I knew the price many of them would pay. And as prey species, sheep are born knowing the score. At least they know us for what we are, predators. They regard people with caution because they live with the bargain their species made with ours ten thousand years ago. In return for our daily care and protection, some of them will die.

Stockmen nurture and kill; I'd have to endure that paradox. Every year, most of the male lambs born in our pastures were destined for slaughter. I was keeping ewe lambs as replacements, but standards are much higher for ram lambs. Anyway, only a few males are needed; their brothers will die.

That night, lying in bed, I wondered how to make peace with my role in the death of livestock. The number of farmers I'd met who shared this ambivalence—who, like me, were animal lovers as children and who found themselves in a more complex role—had surprised me. But if we were farmers, not pet keepers, rigor applied. The need for income, the imperatives of selective breeding, and the desire to produce one's own food meant that lambs would be harvested like ripe corn to sustain human life.

It seemed to me that I should formally recognize the magnitude of our animals' sacrifice. Such a discipline is most needed when something big, something we sense we should be humble before, becomes just another task. I wasn't sure I could ask for forgiveness from my sacrificial lambs and cull ewes, any more than could a coyote. But I'd develop my own prayer, an apology. "I'm sorry, but you must die for the flock," I'd tell them. "For your mothers and sisters, so the flock can go on."

On the hilltop, in early April 2002, I bent over a ewe with my sheath knife and cut into a lump behind her left front leg. I squeezed green pus with the consistency of Cheez Whiz into a yellow plastic margarine container.

Cheesy gland was back.

The stress of pregnancy had triggered a new crop of boils. Days later, I caught two more ewes, performed my crude surgery, flushed their wounds with iodine, and disposed of the puree in our burn pit. I wore gloves and turned my face from the awful craters I was making in their sleek bodies. As unpleasant and upsetting as dealing with CL was for me, the sheep didn't bleed or seem to feel pain as I sliced across the thin, sparsely haired gray lumps. The ewes had walled off the infection: the interior walls of the growths had the stiff, fibrous consistency of a tennis ball.

I'd cull the three after I'd weaned their lambs in late summer. What culling meant in this case was healing their wounds and then shipping them to market. Last fall I had shipped, with our lambs, four ewes, glossy-coated and fat, the disease in their lymphatic systems in remission. I knew their CL would return to infect other sheep if lesions appeared and ruptured before I could operate again. I'd felt queasy sending them into the food chain, though their meat would be untainted. They weren't infectious.

As for the ewes still wandering our pastures without symptoms, there was no convenient test for CL. I was on my own. I'd have to wait until it showed itself by the boils or by signs of the internal form. Last year one ewe was thin all summer, and by fall she'd appeared to be starving, so I'd hauled her to a federal laboratory near Columbus for euthanasia and necropsy. The pathology report said her body was riddled with CL tumors, one the size of a volleyball.

Now I wouldn't test any more thin ewes, just shoot and compost any that didn't fatten after they'd weaned their lambs. So far I'd operated on one ewe fewer than last year, and saw no suspiciously thin ones. I clung to these hopeful signs. And having attacked CL before lambing, I'd prevented the bacteria being shed onto nursing lambs, corrals, and pastures.

Busy the previous fall consolidating the flock on the hilltop, I'd turned in the rams late, so I had plenty of time to watch for the disease. Our first lambs weren't due for almost a month.

One dewy morning at the end of April, I found The Good Mother collapsed in the pasture as if in labor. I kneeled beside her in the grass and touched her nappy white coat, which was like a teddy bear's short fur. Eyes closed, she panted and strained; her rectum protruded. There was no way she could be due to lamb for two more weeks, at the soonest. *This was just like what had happened to Red!* No one I'd consulted and nothing I'd read could explain this ailment in late pregnancy. The problem had to be addressed through shipping Jo's ewes, eating them ourselves, or selling them to Muslim students, and continuing to bring in the genes of new rams.

But The Good Mother was suffering *now*. She had to be helped *now*. Actually she had to be dealt with. There was only one thing to do, I knew by then. I got my .22 rifle and shot her in the head. My stoicism felt familiar: the farmer did this, he who killed, not Richard the kind shepherd. "She was my best hope," I told Kathy as I stood at the sink and washed my hands. While not as tame as Red, The Good Mother had been a larger and more beautiful ewe. Her devotion to her lambs that had struck me when she was a yearling had continued each year, and I'd eagerly saved her daughters.

There wasn't any doubt that our first flock harbored a serious genetic defect, but it was unclear how many of the ewes had it—I'd have to suspect them all. The second flock was host to a disgusting disease, and that foe, too, would reveal its victims in time. Once either problem showed itself, I could take action, but not until. I'd have to wait. I'd been impatient, and now I was at the mercy of time.

Sheep wallowed through our pastures like laden ships through a heavy sea. We celebrated Claire's sixteenth birthday on May 5 as their due date approached, but the ewes seemed too big to wait. At least there was plenty of feed, the overgrown fields producing so abundantly I didn't even consider feeding grain as insurance against toxemia. I couldn't move the flock fast enough to keep up with the grass.

During that warm, late spring, as a flood of frisky lambs finally began to arrive, my hopes rose. I'd vaccinated their pregnant mothers against CL, so the lambs were born with some immunity. And I'd ordered fresh vaccine so that Sam Norton and I could inoculate lambs twice.

I was planning our second hay cut off the former cornfield, but avoided what lay beyond its dark backdrop of trees: the shady inner courtyard and the tree-dotted pastures that I loved. Mossy Dell waited. The old farm slumbered through golden afternoons and the black-and-white mysteries of moonlit nights, paused in its long story of hosting human dreams, fitful dreams that came and went over the decades.

At lambing a year later, our fifth and the second on the hilltop, I didn't wether ram lambs with my chromed rubber-band application pliers as Glen Fletcher had taught me. Although rams would have to be weaned and removed from the flock by three months, to prevent unwanted pregnancies, they would grow faster than wethers and stay lean. And a few could be saved for breeding if they grew well and were especially meaty.

Breeding stock might become my mainstay. Last year it had cost us almost $17,000 for me to bring in just under $10,000 in sales, mostly for meat, for a net farm loss of $7,230.08. The expenses had been so high because I had trenched in water lines and bought supplies for permanent internal fences so that, one day, I could spend less time moving Electronet. I'd been thrilled to note that my "sheep only" costs totaled $5,500; I might have earned $4,240 without the infrastructure investments. As Kathy replied, "*Might have* and *almost* aren't business categories."

The birthing defects in Jo's ewes and the disease in the Goss flock had put a crimp in my ability to sell breeding stock in good conscience. But I thought I'd made progress, or that my luck was changing. And there would be no birthing problems, not one—and no cheesy gland. In early spring I had held my breath, not wanting to hope. Yet by late gestation, pregnancy hadn't triggered any boils. And now, midway through lambing, all the ewes looked healthy, including the yearlings that had been born and raised the year before on the hilltop in the newly merged flocks.

One day I saw Freckles, the little ewe Kathy had picked as a lamb in the Goss barn, standing above her three newborn lambs, which were white with red flecks. Her lowered head came halfway up when I moved in, but she didn't bolt. I stopped short and fell to my knees in the grass. The two bigger lambs jumped up. With a blue steel crook extending my reach by six feet, I hooked one's hind leg and pulled him into my lap. Freckles started to move and I caught the other. She stopped, uncertain, and then turned away; the third lamb leaped toward her. I snagged it and dragged it to me. The two lambs that sprawled in my lap were calm, boneless, but the little one yelled as I freed its leg. I held him against my shins with my left hand. Triplets were a mess to juggle. I had to detain them together to avoid chasing down one or two lambs.

Freckles came up, uttering a soft bleat: "Buuhnnn, buuhnnn, buuhnnn." The little lamb squirmed under my grip and cried as it stretched its neck to her. I knew that to Freckles her triplets had disappeared: sheep, with their eyes on the sides of their heads, not only can't see directly in front of them, they lack depth perception. Freckles could hear her babies, and smell them, but she couldn't see them against my body.

On my belt, in a leather pouch meant for roofer's nails, were white plastic tags and my silver applicator pliers. I punched a tag through each lamb's right ear, avoiding veins. Then I held my notebook on my left knee and reached over the two big lambs—pinning them with my forearm while watching the little lamb—and scribbled down numbers and sexes. Three rams.

I looked at Freckles, who, as always, was tolerating my work calmly. I wrote down her mothering score: +4, the highest possible, as always. I'd sell frustrating mothers first, and so during lambing would exhort myself in my pocket notebook, having learned that I'd forget how upset a ewe had made me as I chased her over the wet grass with a lamb bouncing in my arms. Those chunky notebooks also revealed patterns. Their scrawled data, cohering into meaning as the years passed, would help me breed my way out of woes, genetic and infectious, and toward something better.

But what? Fewer disasters seemed enough that lambing of 2003.

I weighed Freckles's lambs, holding each one aloft in turn in a blue mesh nylon sling from the hook on a blinking digital scale. The two big ones weighed seven and a half pounds, and the smaller one was seven. These were nice triplets. Before the Goss ewes, a newborn twin wasn't quite eight pounds.

I released the trio and they bounded to Freckles. The two bigger ones aimed for her udder and the scrappy little one darted for her lowered muzzle and its comforting grunts. Freckles had matured as the smallest of Kathy's group, a compact white ewe with red hocks and a spray of black spots across her nose. Though docile and sweet-looking, she wasn't pushy or flirtatious as Red had been, and not gorgeous like some members of her cohort. Freckles seemed a wallflower.

But she ate constantly to make milk, and grew a big belly: good rumen capacity for digesting forage. When she'd catch my eye, I'd behold a dumpy ewe, her obedient brood peeking at me from her far side. As with all our original ewes, her ancestry was a mystery, and I wondered why Kathy had picked her. What had she seen in the smallest lamb gazing up at her as Glen Fletcher stood nearby with folded arms? Kathy couldn't remember. Yet probably even then there had been something about Freckles, a calm self-sufficiency.

Then there was Old Mama. She wasn't overly large—and I'd bought *all* the big Goss ewes—but possessed matronly beauty. Her stocky body, a perfectly square block, was covered with short white hair splashed with tan flecks, and borne on flinty black hooves as neat as slippers. Long lashes shaded her watchful eyes; her mobile black muzzle appeared simian, and her lips turned upward in what looked like perpetual amusement. I was thrilled when she also delivered triplets one morning.

As I sat cross-legged in the grass, struggling to tag and weigh her brood, she walked close and began to butt me in my chest with short, sharp punches. Her forehead felt like a pillar of stone. Surprised, I laughed—no ewe had done that, ever—but Old Mama weighed 160 pounds and her jabs stung. When she ran around behind me to search there, I got worried. What if she hit me in my sore spine? Or got a running start and smashed my neck?

Juggling her three lambs, I tried to pivot and face her. I couldn't do much more than turn my torso and try to peer over my shoulder. Old Mama kept circling as I twisted, blasting *baas* into my ears. Growing frantic myself, I forgot to use my most powerful tool, my own voice. Animals respond to soothing tones, something no predator provides. When she turned away, her ears up, baaing into the pasture as if ready to run and search there, I held out a lamb to her—"*See*, here he is!"—but she couldn't register that one of her offspring had floated aloft.

Or she'd lost her mind, unable to accept what her senses told her: I'd swallowed her babies.

It occurred to me that Old Mama surely was one of Mister George's many daughters in the Goss flock—at least she looked just like him. She was a reminder that I'd sold George. Before my exodus from Mossy Dell, eager to make sales, and concerned by George's arthritic shoulder, I'd sold him to a woman who'd called me from Lancaster, just up the road. She'd wanted a ram and two ewes to graze around her husband's milking parlor, and hadn't cared that the old guy limped. I'd made a practical farmer's decision by selling George, but now missed him. And I knew Claire and Tom did, too. He'd been our first sheep. Thankfully our new ram, Kansas, was a fertile sire. And the ram Glen had picked out on our trip had grown into a long, sleek animal. Claire called him Gwin.

I was too busy to dwell on George's absence from the ram pen. But whenever Old Mama cruised past me with a haughty glance, I suspected I'd been hasty in selling a ram that might sire more mothers like her.

I grazed fifty ewes and their ninety lambs that summer on only sixteen acres of pasture on the hilltop. This was huge production for such a small farm. Although I'd been discouraged by sheep ailments, and my own, my passion for pastures might pull me through. I was beginning to know what I was doing as a grazier. Each day, I moved the sheep to a fresh paddock; they even grazed our backyard. With so many sheep cropping the grass, I'd finally been able to keep the spring flush of vegetation under control without mowing. That summer we got nice rains every week, and tender white clover, a sheep's best food, dappled the fields in lush bright green drifts.

With the flock always around our house atop the high central promontory, we had only to walk into the dusk to watch the ewes and their hordes of lambs below on the farm's lesser hills and swales. When the flock had been at Mossy Dell, I was usually headed for home before the magic of evening on a farm, with the chores done and the animals and the shepherd at peace. Now, preparing dinner, we heard wayward lambs crying and their mothers calling. Lying in bed, we heard the stray bleat, followed by a ewe's grunt. Some ewes from the Goss flock wore brass bells—Glen thought they spooked coyotes—and their tinkling was a dreamlike melody.

Best of all were the twilight lamb races. I'd call Kathy and the kids into the yard to watch the spectacle of lambs chasing each other, leaving their mothers and merging like schools of fish. They'd surge up the slopes, only to reverse upon some whim or unseen signal and plunge downward. Stampeding across a flat stretch, they'd swerve and change directions. Flower, ostensibly guarding them, looked bewildered. An animal behaviorist would say the lambs were establishing a pecking order and learning social skills. Their frisky abandon was now enough for me, sufficient proof of the joy of just living.

I'd slept late, done morning chores, kenneled Doty and Jack in the open-fronted shed, eaten a bowl of Cheerios, and was sitting in Fred's former trophy room, drinking strong coffee at the antique oak table Kathy had refinished years ago. Light pressed in from the double-glazed windows we'd added. I felt shaky and had a dull headache from my allergy to the pollen that hazed the air now in late June. Dropping two Alka-Seltzer wafers in a glass of water, I pulled from my pile of reading material beside my place mat the press's latest book by Ellen Bromfield Geld, a memoir about her fifty years of farming in Brazil.

There was a loud knock at the door—Sam. A full half hour early. I liked to move slowly on Fridays, especially during peak pollen season, take time to limber up. No matter what time I told him to come, he was always early—not because he'd earn more; he never even kept track of his time. I did, and paid him ten dollars an hour, two dollars above the going rate. Sam just loved to work.

I took a last sip of Alka-Seltzer and chased it with a gulp of coffee. Without Sam, I'd probably lie around feeling cursed. I sighed and rose, shuffled in my thick socks down the cool dim hall, grabbed my boots beside the front door. Outside there was too much motion, the air warm and moving; even under the porch, the day's brightness felt excessive. Spiky clumps of fescue needed mowing in the too-wide border of grass Fred had left between the driveway and the northern pasture fence. Early was over by seven, and the day felt lost.

"How's Richard this morning?" Sam chirped. He held a white plastic sack that bulged with vegetables. He stood with elbows thrown out, ready to spring into action.

"Sit down a minute," I said, settling into one of Kathy's rustic wooden chairs to put on my boots. Maybe if Sam and I knocked off early, I'd grill myself and Tom lamb chops, and then we could go hit golf balls. Kathy was on a road trip with Claire, visiting prospective colleges; they were in Wisconsin, at Kathy's undergraduate alma mater, and tomorrow they'd head for Minnesota.

Already it was hot, the breeze gaining confidence from the southeast. Sam and I stared out from the hilltop, across the dusty white gravel driveway and over the low northern pasture to Ernie's scruffy hill beyond. On the rise just before the road disappeared, Ernie's discarded farm equipment, cinnamon with rust, settled into weeds at the base of his abandoned silo.

"Ernie's wife, Janet, is sick," I said. "Kathy's been making her chicken and dumplings."

Sam took this in and offered his own intelligence: "Did you know Fred's sick? Prostrate cancer is what I heard."

"*Prostate* cancer won't kill him."

"He's been coming to our church." Sam gave me that sly look of his, knowing how I felt about Fred, and pitched his voice high: "Oh, he got religion. Yes sir."

"Scamming someone new," I said. "The almighty Lord!"

I worried I'd gone too far, blasphemed, but Sam looked distracted. Sitting in a hickory chair that matched mine, his forearms and feet planted and pointing straight ahead, his hands clasped and released the rounded ends of his barky armrests. He had new Vibram soles, fishbelly white, on his ancient leather boots; he credited their longevity to the used motor oil he slathered on them. I doubted the worth of his Quaker State applications— surely that oversoftened the leather—but I had to buy a new pair every other year and listen to Sam brag about how his footwear had carried him through another season.

He stood and swiveled toward me, braced. "Let's do something even if it is wrong," he said.

That was always his final utterance, used to hurry me along. No more goofing off. Sam had nagged me about cleaning up the barn until I'd agreed. We'd work all day tomorrow, too, unless I made some excuse for quitting early, such as needing to run Tom into town. And then Sam would continue during the week while I was at the office.

We passed from summer's insistent light into the barn's timeless shade. From a pile of broken square bales, Sam collected loose hay on his pitchfork's tines, and hustled across the aisle to throw it in the feed bunk. I joined in the dusty work. On the other side of the partition, Kansas and Gwin stretched and walked up to sniff the feed. Particles rose and hung, an amber mist in the rear doorway's light. My nose clogged and veins throbbed at my temples. Certain I'd feel awful tomorrow, it irritated me to know that Sam would show up early again, eager to start cleaning out the ram pen.

I stabbed a pile of fodder with my pitchfork, took three steps, and heaved it at the manger.

"My new hay is going to be even worse than this crap," I said.

"How's that?"

"Cut too late."

Sam halted just before loosing a mighty fork thrust, turned his head to look at me, and slowly uncoiled from his tense crouch; his pitchfork's tines lowered, one of his hands rose from its rubbed handle, and he scratched at his cropped bristle of white mustache.

"It was too rainy to get an early cut," he said. "Folks are baling now."

"I know. But it's too late for decent quality. And I can't get anyone to bale mine until they've baled theirs. I might as well feed cardboard."

"It's better than eating snowballs," he said with a snap of his chin, and wheeled to gather another bundle.

That was what Sam always proclaimed whenever I bitched about my winter feed. I'd heard other men here use the phrase, the response generations of Appalachian farmers had distilled to defend their lousy hay against ambition. This wasn't alfalfa country, and tough Kentucky 31 fescue crowded out better forages anyway. So they did the best they could; there was plenty of other work, and the livestock would winter with something in their bellies.

I suspected the fate of Mossy Dell lay behind my anger over the hay. Before leaving with Claire, Kathy had finally broached what was on our minds about the valley farm: "Maybe we should think about selling it." I had been trying not to. Yet we could replenish Claire's and Tom's college funds. Mossy Dell was an expensive resource to pay taxes on, mow, lime, and fertilize just for a landowner's annual 50 percent share of coarse fodder. I could buy better hay cheaper. And I still had no plans for returning sheep there. My headlong charge had halted.

I'd agreed with Kathy and felt relief, which instantly made me feel traitorous. To what? I wondered. The old farm, an unconscious parcel of land? A hazy plan to run 125 ewes there? A dream that hadn't panned out? Anyone could see what we should do; I just didn't want to do it. And in this I glimpsed, like a rat in the shadows, my passivity. My farming had something to do with fighting this flaw in myself. With doing, not just dreaming.

Ending Mossy Dell's drift, and ours, was one of life's unpleasant adult tasks. Kathy had been reticent only because she knew how much I loved that farm. Or was it "had loved?" Given my lifelong desire for a farm that beautiful, that didn't seem possible.

CHAPTER FIFTEEN

At the start of lambing three inches of rain fell. By afternoon it was sleeting, which changed to snow in the night. We're warming up soaked lambs by the woodstove.

—e-mail to a friend

LONG AGO I'D STOPPED TALKING OF BUILDING A NEW HOUSE AT THE valley farm and selling the hilltop. We didn't have the physical or emotional reserves to remount that rollercoaster, money concerns aside. We'd become averse to disruption. Our house, new from the ground up, reflected Fred's legacy in the form of a wet basement. But it was home. And it didn't appear that I'd ever run sheep on both places; I didn't have the energy for it, and feared getting hurt again. Other than arranging hay cuts on the roadside field, I'd hardly visited Mossy Dell in two and a half years. There was no reason to; in retrospect, it was too painful.

So in March 2004 we sold Mossy Dell, our beautiful farm, my boyhood dream, our shared wish.

I'd always assumed that our selling would result in the place being split into plots for houses by its new owner. After we first listed the land, developers did look at it. We frustrated our realtor with our lack of enthusiasm and our limited desire to bargain, especially with anyone who wanted to subdivide. We refused two low offers, didn't mourn when several other prospects drifted away, and finally took the farm off the market.

Then, one night in late winter, a woman called. When she was young, she said, she'd been close friends with Kenneth and Mabel Vaught. Like them, she loved horses, and her husband loved hunting and country life. And as it happened, their daughter was about to marry and would need a home. We closed the deal without much haggling. We might have made more money by listing it again, but in the end, the right people got Mossy Dell.

The newlyweds did what we should've done: they bought a prefabricated house, which arrived in early summer on flatbed trucks that convoyed down Ridge Road and made groaning right-hand turns onto Snowden Road at Sam Norton's house, also factory built. Such homes were popular in our area because they made financial sense, eliminating confusing squads of subcontractors. My parents' house at Coral Tree Farm had been prebuilt, and assembled onsite in about a day—the perfect time frame for my impatient father.

I'd been certain that Mossy Dell's new owners would raze Massey's shed in the north pasture and put their house there, overlooking Lake Snowden. Instead they built at the entrance to the south pasture, east of the old cabin site, above the placid farmyard pond that I'd rebuilt with such trauma during a dry year five years before. Sam told me about it; he said it was large and attractive. I never drove over to see it, not trusting my emotions enough to assume I could risk a casual visit.

The sale of Mossy Dell more than replenished our college savings for Claire and Tom. For all of my mistakes, the old farm had turned out to be a good investment. Just before we closed the sale, I explained this to my mother. If we could recoup money and earn a bit, we could chalk up the whole misadventure to experience. We could rectify my overreaching.

"But it's the end of a dream," I added.

"So what?" she said. "You still have a farm."

"A small one."

"That's all you have time for."

Her bluntness startled me, destroying the fragile calculus between head and heart I'd just offered in vague hopes of commiseration. Confused, I ended our talk. I wanted to avoid her rubbing my tender romantic nose even more vigorously in the Facts of Life. She was an expert on dreamers, after all. And I'd adopted her view that we'd have been wealthy if Dad hadn't single-mindedly chased his passions. He might even have succeeded as a farmer if he'd heeded her cautions about spending.

Yet I believed in Dad, in a way that departed from Mom's mythos, in a way that maybe only a son like me could. Not only had I adopted his agrarian interests, I respected what seemed to me the virtuous seed of his dreams: loving your work for more than what it paid. Mom believed that a good income bought a good life. Though Dad liked making money, most ways of earning it didn't interest him. He'd explored at various points self-employed careers other than farming. In his library were books on cartooning, freelance writing, commercial fishing, and plumbing. At the Georgia ranch he'd taken correspondence courses to become certified as a physical therapist. I'd found this dumbfounding when I'd struggled in the early years of my own farming. Mom, pregnant with Pete, had been Dad's practice dummy on our dining-room table. He never worked as a therapist, and his physical modesty made that profession unthinkable to me.

In the hard choice between his money or his life, he'd finally chosen money—to support his family—though it was difficult to picture Dad just scraping by, at anything, on love alone. Confined to an office after his brief ranching career, he'd been restless, resigned, and successful. Working in a hierarchy, beholden to others and answerable to a boss, had disciplined his impatience and awakened his heroic impulse to sacrifice himself serving a group. At last, as a consultant, Dad had opted out of the beehive again while serving its minions in a quasi-independent way. He'd always been a great organization man. As Mom put it, "He can make money for someone else but not for himself."

Wild ducks still appeared on our hilltop. They visited our pond, eyeing me warily as I moved electric fences, and ventured inside the barn to eat with the chickens and guineas from a trough inside the big open doorway. Sometimes a pair lingered nearby, as if trying to remember something. The green-headed drakes and the stippled brown hens summoned memories of

Claire and Tom cuddling seven downy ducklings. Our visitors brought to mind a secret nest, a dry year, and a ferocious thunderstorm.

Claire and Tom and I had hatched the original mallards in our living room; they'd emerged already imprinted on my voice. The kids held and petted them; yet they were wild creatures, and the marshes had called to them. First they'd left the farmyard and made their way to the pond in the hilltop's south pasture, and then they crossed to the lake. I had raised others, maybe their descendants, from a clutch of eggs I'd uncovered in the barn's haymow and from a nest I'd found in the remnants of a round bale in the pasture below our house. Nature wants reproduction above all else on this messy planet, for a mated pair to replace themselves before dying, and I'd done my share to help Lake Snowden's mallards with their most important task.

I liked to watch the ducks wing back to the lake, crossing above the trees that overhung Mossy Dell's farmyard. The waterfowl carried me back to the old farm, into the farthest reaches of Lake Snowden and beyond, into the wholeness they inhabited. Anyone who raises a flock of wild ducks learns, the first time they lift into the air, that they carry a piece of your heart with them into the sky. You can feel it go—a sudden fear, so unexpected—and then a thrill, a joyous unfolding.

That should've prepared Kathy and me, but didn't, for seeing Claire drive away alone the summer before for the first time. I'd spent hours practice-driving with her on the campground lanes around Lake Snowden. She'd gotten her adult license; she was ready to fly. We'd walked outside behind her that Saturday afternoon, to our gravel parking spot across from the house. "Be *careful*," Kathy said as Claire got in the car and stuck her key in its ignition. Claire reached out to pull her door closed, and then she tossed back her hair. She looked at Kathy. "I *will*, Mom. Don't worry!"

We watched the little red car we'd bought her, a used Toyota Camry, roll slowly away. Claire aimed herself toward Ridge Road, gathering speed, leaving Kathy and me standing alone together on the hilltop. Claire's cats rubbed against our legs. Kathy's hand flew to her mouth; a strangled sob escaped her throat. "Wait," I wanted to yell, "I'm not ready for this."

But Claire was gone, happy behind the wheel, disappearing down the driveway and into the world.

One afternoon that July, a woman upset Gwin, our gentlest ram—but 240 pounds—by petting and pressing on the top of his head. She was leaning over the homemade wooden hurdle I was using as a gate to his pen. I'd been talking to her husband in the barn's doorway when I glimpsed Gwin lowering his head and backing up, preparing to charge and bash her through the slats. She turned, her forearms against her chest, and fluttered her hands in the air at me. I hurried over to where she stood against the barrier. At my approach, Gwin relaxed and walked up.

None too gently, I said, "Rams understand only three things. Feeding, fighting, and . . . breeding. If you touch the top of his head, you're challenging him to fight." I showed her how to hold him under his chin, or scratch behind his ears, to show affection. "But," I repeated, "don't touch the top of his head." She giggled. Some people couldn't see that humans and animals converse without words, a dialogue that had become second nature to me.

I could bristle about what city people "knew" about sheep—their stupidity and meek passivity—while ignorant of their humble glory. No good farmer has contempt for his animals, which daily instruct him with their simple presence, their stoic acceptance, their fleeting displays of joy. A freelance book designer, on a visit to our farm that summer with her husband, curled her lip with knowing distaste as she asked, "Aren't sheep stupid?" The sly bigotry of her facial expression startled me. I thought of Freckles and Old Mama. *How smart do you require the animals you eat to be?* I wondered. "They're smart enough," I said, "to be successful sheep."

I had a bias for my chosen species, but I wasn't being paid to serve as an image booster for sheep. I was trying to make money—well, at least break even—as a farmer.

Which was why it seemed ironic, even to me, that I'd been spending my lunch hours at the press in my truck, parked in the shady lee of a campus sycamore grove, hand-drafting on a yellow legal pad a story for the *Stockman Grassfarmer*. Busy with publishing and farming, I hadn't written a freelance article in years. I'd pitched the magazine the idea of my profiling Doug and Laura Fortmeyer, the breeders of our rams; the frugal grass farmers in Kansas never propped up their ruminants by feeding grain, and they selected for mothering ability on pasture. The *Grassfarmer*'s editor had agreed to pay me $200. I'd count it as farm income.

Kathy was thankful that my farming at least gave us a tax write-off. And I was eager to show I was trying to profit. I'd already received several deposits for ewe lambs and had many ram lambs to sell for meat; income would be about the same as the past two years, $10,000, but so would expenses. My only cost that was running below average was for labor; Sam hadn't been feeling well, and Erma was taking him to doctors for tests.

Infrastructure would be costly again. I was pricing an expensive new sheep-handling system—$5,000 for galvanized chutes and gates—that would help me sort, weigh, and deworm sheep when Claire and Tom were both at college. And I'd already bought a new chain saw for $280, spent $500 to replace my truck's tires, and shelled out $1,000 for fresh gravel for the driveway and the farmyard. Such purchases were necessary just for living in the Appalachian countryside, so it was nice that the farm subsidized them.

Yet it bugged me that, like my father, I couldn't seem to control costs. I'd managed to spend $1,000 every year on vaccines, drench, and veterinary bills for a flock of only fifty ewes and their lambs. The *Grassfarmer* article was the easiest farm money I'd earn this year. I did chores every day in all weather, ran a business, and managed a complex ecosystem. But I'd have pocketed more money as a part-time greeter at Wal-Mart.

"Well, do you feel ready?" Kathy asked Claire.

"*Yes*. Mom. I'm ready."

"You can get your books tomorrow."

Claire nodded once, sharply.

"Maybe," I said, "you and your roommate can go get ice cream tonight."

"Yeah."

"She seems nice," I added. "You love musicals, and she wants to act in them."

"There's lots of fun social activities this week," Kathy said.

"*Dad? Mom?* You two need to chill." Claire widened her brown eyes for emphasis: we were annoying.

I looked at her across the table; she appeared small, pale, thin. Sensitive to where meat comes from, she'd turned vegetarian in high school. She'd stuck a PETA button on our refrigerator; the slogan, under a photo of a fluffy baby chick, said, "Your food had a face." Now, hunched forward, picking at her small salad, her long brown hair on her shoulders, her elbows tucked against her ribs, Claire seemed gathered into herself, tensed for a leap. This was her next-to-last moment with us before we left her at Northwestern University. She'd already said goodbye to Tom, who was staying with a friend in Athens. Kathy and I had hoped to linger with her over a delicious meal. But I could barely hear her replies above the clatter in the "Asian fusion" restaurant we'd picked. The tofu dish I ordered was awful, so I was eating Kathy's meal—dark bits of beef, tasty but too salty.

I hadn't gone on any college visits with Claire, and wanted to share this trip. And to redeem myself for also missing her high school graduation in June. Even though Claire hadn't wanted to attend her own commencement, my absence had seemed indicative to her of my misplaced priorities. For once my day job, not our farm, had been the reason. When she got her diploma I was in downtown Chicago, not far from where we were dining, at BookExpo America, the world's largest exhibit for book publishers. Promoted the previous fall to marketing manager of Ohio University Press, I was responsible for selling $1.25 million in books a year by promoting our backlist and fifty-five new releases.

At BookExpo I'd had trouble eating, too. I'd spent every day, in the booth that I'd secured, furnished, and stocked, with an ache in my lower abdomen. Finally I'd called my doctor in Athens. "That sounds like diverticulosis," he said. "An intestinal inflammation. You probably haven't been drinking enough fluids." Gallons of strong coffee had gotten me through my longer office days, but apparently had dehydrated me. Yet I'd always trace the pain to guilt over missing Claire's ceremony rather than to anxiety over my new duties.

To allow me to come on this trip, Kathy had found a farm sitter. Sam had gotten sicker—it was cancer. He'd telephoned one night recently to tell me, and then handed the phone to Erma. "I can't say the name," she'd said. "I wrote it down. The doctors haven't seen it before." My Google search later returned only two hits for the rare blood cancer, and almost no details. What worried me and Kathy as much as Sam's sketchy diagnosis was that Erma had been taking him to doctors in Jackson, a cow town thirty minutes west. For cancer we'd at least go to Athens, which seemed risky enough—two friends had nearly died in the town's hospital from doctors' mistakes—or up to Lancaster or on to Columbus. I wasn't sure Sam and Erma believed in doctors, and maybe that was why they thought one was as good as another.

"Where's that Barnes and Noble?" I asked as we left the restaurant. "We could walk back to campus and stop in. Do you want to, Claire?" She hesitated and then said, "I have a meeting. Remember?"

As Kathy drove us across town, I gaped like a rube at suburban Evanston, drenched in late August light, so reminiscent of the prosperous Indiana town we'd left eight years before. Lofty oaks, maples, and honey locusts shaded gracious Victorian mansions and dotted Northwestern's parklike campus; the dark trees seemed to draw sustenance from the blue waters of Lake Michigan that lapped the nearby shore.

"What a great place to walk this would be," I said. "Like Bloomington. Winter wouldn't be so great. Can you imagine? Two feet of snow, an icy wind off the lake."

"That reminds me, Claire," Kathy said. "Did we pack your boots?"

"Yes. *Mom.*"

Outside Claire's dormitory we perched on a bench. Coneflowers hung in the warm air around us like pink shuttlecocks; a fat bumblebee clung to the brown button eye of one wavering blossom. Kathy reviewed the use of debit cards and fumbled a speech about making the most of one's college years. Claire glanced toward her stone dormitory. "Kathy," I said, "if we don't leave, she can't miss us." Kathy patted Claire's shoulder. "Call us," she said, turning away as her face swelled. She looked in her purse for a tissue. Claire stared at Kathy's lowered head and threw out her arms in theatrical frustration. Parental emotion was too heavy to carry into her new life.

Kathy and I returned to our hotel, about fifteen miles from Evanston in

Oak Park, where we had slept last night in a double bed with Claire beside us in another. In an oddly timed coincidence, a publishing conference based at the hotel had dovetailed with Claire's campus move-in. I'd spent the day here in meetings while Kathy got her situated. As we entered our dim, cluttered room I noticed Kathy's annoyance: I'd left a Do Not Disturb sign on the door, so our beds were unmade and damp towels littered the bathroom floor. She went to brush her teeth, and I kicked off my shoes and sprawled on our bed to read.

From the nightstand I picked up bound galleys of a book we'd publish that fall under our Swallow Press imprint, Gene Logsdon's *All Flesh Is Grass: The Pleasures and Promise of Pasture Farming*. I'd courted Gene when I was at Indiana University Press, which later published his memoir. And while Gene's first book for Ohio University Press, the fable about the man who heals strip-mined Appalachian farms, had sold poorly, I was sure *All Flesh Is Grass* would join *The Sheep Book* as one of our bestsellers.

I dog-eared the four pages where Gene had quoted me—as a farmer, not as a publisher—including from my e-mail to him about my triumph growing lespedeza on the abused cornfield. The book's cover was a color photograph of Gene's sheep grazing with a red Hereford on an emerald sward; its contents featured happy pastoral homesteaders, their fields as diverse and succulent as salads. *All Flesh Is Grass* carried a hint of Louis Bromfield's romance and was free of the foment over income that permeated the *Grassfarmer*. Yet I'd imbibed the magazine's credo for so long and studied literature even more technical—and struggled so hard in our own fields—that I found myself skimming. Gene's news was for novices. "It's pasture porn," the man I was training to do my old publicist job liked to say. I always laughed at his jest; I always felt a tad defensive. *All Flesh Is Grass* was manna for someone I knew, some lost boy in a sterile mall bookstore looking for visions of an agrarian Eden.

I kept turning back to read Gene's dedication: "To Richard Gilbert."

Kathy left the bathroom in her nightgown and climbed into Claire's bed.

"You aren't going to sleep over here? With me?" I asked.

"No, you're reading. I'm tired."

After a while I felt sleepy myself and decided to put on my pajamas. I slid to the edge of the bed and placed my feet on the floor, but when I tried to rise, my knees buckled. I gripped the quilted bedspread with both hands. *Wow, dizzy*, I thought. But I wasn't dizzy. My head was just fuzzy, like I was having an allergy attack. I launched myself toward the bathroom again, and started to fall. My leg bones were jelly. I collapsed backwards onto the bed.

"Kathy?" No answer. I heard whimpering. She was clutching Claire's pillow with both hands and had her face buried in it.

"Kathy?" I tried again. "Something's wrong."

"We've lost our daughter." Her voice was muffled, but she was crying.

"No. I mean . . . *I can't walk.*"

"You don't need to walk!" She was crying harder.

"This is serious. I'm sick. I can't even get to the bathroom."

"You don't need to go to the bathroom!" she wailed.

I lay on my back and thought. *Kathy's dinner, that beef I ate.* It must've been full of monosodium glutamate, which had never bothered me but which my brother David was very allergic to.

"I think it's MSG poisoning," I announced. "Do we have any Benadryl?"

"Just go to sleep!" she howled. I heard thrashing over there. I wondered, as my eyes closed, if I'd wet the bed tonight like I used to do as a kid.

Kathy awakened me. She was dressed, holding a glass of water in one hand; two pink pills rested in the palm of the other. "Take these," she said, leaning over me. "We've got to get on the road."

Soon we were inching in the dark past flashing yellow lights. With the highway narrowed by construction, traffic was snarled even though it was four o'clock. Kathy gripped the steering wheel and ignored my complaints. My legs were working, but I could have used more sleep. The only reason we had to start this early was because she had to get back to campus to teach a class. A real sore point with me—*administrators don't have time to teach*—and I felt neglected when she did. And she wasn't just a dean anymore: two weeks ago she had become provost, the university's chief academic officer, second in command to the president.

At daybreak we lowered our visors as Kathy drove us silently toward a blinding sunrise. Maybe she was upset with me. I was sure angry with her.

And the longer I sat, the sadder I felt. Tom would leave too, in only two years. I felt like I was falling. Down I went, alone. I was afraid I'd hit bottom, and feared that I wouldn't. This chasm was bottomless.

When Claire was tiny, back in Indiana, after Kathy and I had eaten dinner and drunk our glasses of wine, we'd steer her in her little blue canvas stroller through our neighborhood. My Labrador, Tess, who'd helped me court Kathy at Ohio State only four years before, ambled along, her black coat as glossy as a panther's. Kathy's brown hair was long then, and curly, and bounced. One evening in Claire's first summer we went extra far, and up ahead we saw an elderly man slumped in a lawn chair inside his garage, looking out at the street. Our university friends made fun of this very Hoosier habit, lounging in garages, but we secretly liked being the young couple that such geezers smiled at. We waved at him and paused as Tess sniffed the curb. He studied us, his head cast to one side. "You're happier than you know," he croaked.

Now I hated him, that long-dead man whose words rang in my ears. It had seemed true, what he'd said, but even then it felt somehow mean-spirited. *Of course we're happier than we know*, I'd thought, and tried to dismiss his words, but clearly he'd been *us* once. Isn't such wisdom just a curse if you're not ready for it? I wondered now. Like telling a child, "Enjoy your life, kid. Because you're going to die."

I looked at Kathy as she squinted into our first dawn without our daughter. My bleak mood—my regret—felt toxic.

"It's like our purpose is over," I blurted. "Like there's no reason to go on."

"*How can you say that?* Tom's still at home. Claire still needs us. The kids will need us for *years*. We have great jobs. The farm. We have so much to be grateful for."

She was crying. Even a spouse shouldn't have to endure all of her mate's craziness, not all at once. Was I was really just starting to see, so late, that having strong feelings didn't make me special? That they certainly didn't make me good?

In late September I received an e-mail from a man in West Virginia who was looking for a ram. John Stenger's family background was dairying, he said, but he and his two young sons were running beef cattle on five farms, and had a herd of meat goats and a flock of hair sheep. In reply, I bragged about the size of our ewes and the growth rate of their lambs. John was a third-generation farmer from the mountain state, he later told me, and had seen other breeders use the same pitch. He wrote back:

> My top ewe should be the one that gives not the highest production, but more importantly, the highest net return. This is damn near impossible to measure precisely, whereas production and growth can be easily and accurately determined.
>
> Older ewes with proven superior production, health, and longevity, that have produced several excellent daughters and have lambs that grow rapidly to about 100 lbs and then slow down to a moderate mature size, this is the sort of ewe from which I would ideally choose my rams.

I'd never received such a thoughtful e-mail from anyone in the sheep world. I felt chastened by John's gentle critique. After my first chaotic lambing, my goal had been to select for mothering ability; nothing is more important in pastoral livestock farming than for females to be successful mothers outside, on grass. Yet the common desire to breed bigger, impressive-looking animals had infected me.

"You've described Freckles," I replied. "She'd mother a fence post if she gave birth to it, and she's not very big."

In fact I'd wished fervently that she were bigger, though she was one of only two ewes—Old Mama being the other—who seemed comfortable raising triplets. Freckles's triplet litter that spring had weighed twenty-four pounds at birth—nice eight-pound lambs—and when I'd checked them again in sixty days, as I ran lambs through the box scale in the barn, they'd totaled ninety-nine pounds. I still hadn't learned how impressive that growth was for triplets out in the pasture with no grain. And at that age, their growth mostly reflected Freckles's mothering ability.

A big brown ewe in Freckles's cohort—Fancy, I called her—had twinned that spring, and at the same checkpoint her offspring weighed seventy-five pounds. Although by fall her lambs had grown much bigger individually than Freckles's, as befitted their easier life, the total weight of Freckles's brood was still slightly heavier: more pounds to sell. Or to keep to breed more sheep like them. Without records, almost anyone would've picked out Fancy's stockier twins to keep. Even with my lambing book, even with my clipboard filled with lambs' sixty- and ninety-day weights, even knowing that Freckles's daughter was a triplet, it was hard to go against my emotions and pick her scrawnier lamb for the flock. In fact, I hadn't. I had kept one of Fancy's gorgeous twin daughters, and had sold Freckles's lone daughter along with Fancy's other lamb to a man starting a flock. And both of Freckles's sons were in the meat lamb pen.

Having lived so long with Jo's bad decisions, her lack of rigor that had created genetically flawed sheep, my illogic had been vaguely troubling me. It was dawning on me that favoring the physically impressive sheep from the Goss flock might lead to clumsy, indifferent mothers. A shepherd can only emphasize so many traits, and size itself seemed hostile, eventually, to maternal ability. (Fancy, whose maternal skills seemed at best fair, next year would reject a lamb.) John's clarity about his goals—my own first goals of profit and pastoral fitness—made me feel like a confused bumbler. And with my malleable sales pitch, a hypocrite to boot. I'd collected data, unlike many shepherds, but hadn't used it. I had to start acting on my convictions.

In one of my father's articles for a flying magazine, he'd once written that many a pilot has flown his airplane into the ground rather than ignore his emotional perceptions and believe his gauges. I'd been doing the equivalent, I saw.

Late in his life, Dad got new books on writing. He'd liked introducing people to aviation and agriculture, risky realms his readers could enter safely with his clear prose. In his articles, as in *Success Without Soil*, he underscored his own mishaps, presenting himself as a vaguely comic figure. To me, his own heartbreaking childhood was his obvious topic, but I knew he wouldn't excavate that pain. Nor would he view himself as a suitable subject.

Mom and Dad had reconciled two years after he'd sold their nursery. Their new stucco house had a swimming pool and was located in a subdivision called Suntree, two miles from Coral Tree Farm. Mom had moved out of her inland apartment and Dad had sold his beach condo, and they'd met in the middle. Dad told Mom that his first year living at Suntree with her, working as a consultant and tending their lawn and cars, had been the happiest of his life. He said he finally knew what was important. He'd been touched watching handicapped people struggle to stuff envelopes for him, Mom reported. He wanted to help them one day. And perhaps, he told me, he'd learn to play the piano, too.

As always, when Dad talked like a regular person, I didn't know what to think. This was partly because after the original statement, issued like a bulletin from an alien shore, he sank back into silence. I couldn't imagine he would ever truly retire, become a piano-playing community volunteer.

Dad's last Christmas, in 1989, we'd all gathered at their house, which inside was golden and red, sparkling with Mom's decorations, and comforting to us with the smells of her cooking and with the dark Spanish furniture from our childhood. There were grandchildren everywhere. My brother Pete's daughter, Keeley, swam with Claire—the girls, both two years old, scaring themselves with the image of a blue crab inset into the tiles on the pool's bottom. Dad posed for pictures with my brother David's son, Chase, six months old, and with our Tom, only two months old.

I told Dad of my plans for our Indiana homestead and showed him photographs of the pond we'd dug. The only structure on the property was my tractor shed, and Dad asked how his little Kubota was holding up. By then I felt confident in reporting that he was right, that the tractor, turtle-like, would chug forever.

Then Mom and Meg gathered the boys—David, Pete, and me—and told us that Dad's heart was working at only 8 percent capacity. He didn't know it was that bad, they said; the doctor had told him 20 percent, bad enough. To get through airports during his consulting trips, he popped nitroglycerin pills.

He wanted to talk to us now. Dad reclined in his armchair in the living room, his feet up on an ottoman. He said his heart was failing and tried to reassure us. "I'm not in pain," he said. "I'm not suffering like I would be with cancer."

I fell on him, kissed his rough cheek, tried to hug him. He submitted quietly, unmoving, his face slightly turned from our first embrace.

On a Saturday night the following December, Meg called me in Indiana to say Dad had passed out while she and Dad and Mom were at dinner. By that time, his heart was barely pumping blood to his brain. He'd rallied, and Meg put him on the phone. It didn't sink in that surely I was talking with my father for the last time. We'd always had terse phone conversations, me needing more than he could give, frustrated. This time I was overwhelmed by our house construction and distracted by the coming work week.

"You really gave everyone a scare," I said lamely.

"It goes with the territory," he replied, the phrase epitomizing his stoicism and the unselfconscious machismo of his generation.

I was under water when my father died. That's how I got the news, anyway, while submerged. It was early the next Monday morning, and getting ready for work, I'd immersed myself in the bathtub to wash my hair. There wasn't a shower in the rental house where we were living while building. Kathy's hand reached through the water and grabbed my arm.

I surfaced, looked at her. "Your mother just called," she said. "Your father has collapsed and they can't revive him."

She'd told me Dad was dead, but that's not what I heard. This was just another medical emergency. He'd collapsed before. Mom had gone with Dad to the hospital, where they would be able to fix whatever was wrong. He'd died already, of course, on the kitchen floor. She'd heard him make a sound, as if he'd taken a blow, and then he was down.

When I arrived, Mom and I hugged. We tried not to sob, but my shoulders heaved. "It's big," she said, shaking her head and wiping her hazel eyes, standing back to look into mine. She said Pete would take me and David to see Dad. It wasn't allowed, she said, but Pete, as a cop, could get us into the hospital's morgue. It was our last chance to see him. His wishes were to be cremated, and for there to be no ceremony, to vanish without a trace.

After the hospital, frightened by the wildness of my grief, I wrote his obituary for the local newspaper. I stayed up all night and described his passions for aviation and agriculture, of his leadership of thousands of workers at the Kennedy Space Center. I didn't mention how he secretly listened to country music on his truck's radio, which he left on so that it blasted our ears when we borrowed the little red Chevy for errands. I didn't write about his sense of humor, surprisingly silly. Or his humility, his intrinsic morality, or the way, without trying, he exuded authority. I didn't say that he was an exile, a pilot who'd never really landed. Those were my perceptions, surely not close to his interior reality, and anyway, what he'd done with his life was the point.

We gathered the next morning in the living room, Mom and all of his children—her four and his first wife Jean's two, Ann and Chuck. Ann, who'd insisted on some observance, read from the Episcopal Service for the Dead. Meg's husband, a pilot, mentioned Dad's membership in the Quiet Birdmen, a secretive fraternal order that Dad's hero Charles Lindbergh had helped establish. The room resonated with Dad's life, with us, looking at each other. But especially with his wild years in California. There I was, named after his best friend, who'd been William Randolph Hearst's favorite pilot; my brother David was named after another friend, a frequent weekend guest at the desert ranch, Dad's former psychiatrist.

I sobbed in my seat on the airplane going home. My parents had taught me never to cry, and I fought for control. In Indiana, the limo driver let me out onto a heavy snowpack in front of our tiny rental house. Stripped by loss, I walked under the maple trees that stabbed the dark sky beside the icy street, toward yellow lights where my family waited. They knew me as a husband and father, not as a son.

Soon after my return, Mom and Meg and Pete took Dad's remains to Coral Tree Farm. He'd told Mom, one day when they were living there, to scatter his ashes under the oak tree beside the farm's well. The new owner, the developer, who lived there, gave Mom permission.

After everyone left Mom, returning to their separate lives, Dad appeared to her in a dream. He was glowing in the way he could, and said, "I had no idea there was so much to learn." That sounded like Dad and like the conception he would have of heaven. Mom put a wry spin on it: "He didn't ask me how *I* was doing."

One day the developer bulldozed Coral Tree Farm. He flattened the poultry barn, the grape arbor, the giant trees, even the cedar house. There, on the bared sand beside his truck terminal, he paved the farm with concrete and erected an office park. But he spared Dad's oak.

Pete visits the tree when he travels through that part of the county. There's a picnic table beneath it where the office workers take breaks. Nothing looks the same, but Pete walks and thinks about Dad and the farm, remembering the place in surprising detail. In his mind's eye, he sees everything beneath the surface, everything that was obliterated.

John Stenger and two of his sons arrived from West Virginia late one Saturday afternoon, the last day of October, in a pale blue flatbed farm truck with wooden slats enclosing its sides. A man in his mid-fifties with a wild shock of graying sandy hair, unshaven and dressed in rumpled clothes and calf-high brown rubber boots, John fit the image of a rustic Appalachian farmer (though I'd learn later that his father was an environmental scientist with a PhD). His boys were tanned and had thick crops of dark brown hair. Bob was reserved, about twelve, lanky and starting to get his growth; Frank, a little younger, was husky and grinned with pleasure over their outing. The boys were quiet and watchful as we crossed a sloping paddock below our house to see the ewes. John and I talked about breeding as we rambled, and he surprised me by uttering the words "selection pressure," my only customer ever to do so. We were talking about how shepherds whose ewes give birth on pasture must favor maternal skills.

I said, preaching to the choir, "I think grass farmers need a higher standard than those who lamb in the barn and confine ewes in four-by-four jugs with their lambs. How would you know, when a ewe can't get away from her lambs, that she would've mothered them if she'd been on pasture?"

"You can't," John said, "and it takes only two generations to lose it."

When we reached the flock and I pointed to Freckles, John nodded. At

140 pounds, she actually was heavier than what he'd said was his ideal ewe by about 20 pounds; his statement about "efficient" mothers meant they had small bodies that could convert a modest amount of forage into lambs. Lately I'd been lusting after 170-pounders.

We returned to the barn, and John's boys plucked from the lamb pen a wiry white lamb with brown spots and red hocks, one of Freckles's sons sired by Kansas. "How much do you want for him?" John asked. I was selling ram lambs for breeding for $250, but didn't feel I could ask that much for a lamb I had been fattening for slaughter. "He's $150," I said, naming a premium meat price, and John wrote me a check on the hood of his truck. Before he drove down the hill, headed home, he leaned out his window and said, "I'll buy one of her sons every year."

I hurried along the driveway as the mild afternoon edged into evening, to a glass of wine with Kathy before dinner, eager to tell her about my most discerning client, my latest role model and newest hero.

The sale of Mossy Dell made permanent a shift in my farming that had begun the day I lifted Red into my truck. Instead of shepherding a large commercial flock that would fill Mossy Dell's quirky paddocks, flow into the big new pasture created from Fred Paine's cornfield, and cross the road to populate the hilltop farm, I would sell breeding stock.

That tack usually meant a smaller flock, which was all the hilltop could support anyway, but sheep with value added through pedigree records and breeding progress documented by data. My interests had always run toward breed improvement. And my experiences certainly had confirmed the need. We now owned a small farm, barely sixteen acres of pasture, yet it was big enough for an intensively managed seed-stock flock.

An Internet presence would make our acreage large enough. If sales lagged, I'd flog the digital world harder. About half of the people who now contacted me about buying stock or meat did so through e-mails.

The farming niche of the Internet had mushroomed. This domain, full of information and ideas, was suffused with humans' everlasting hunger for community. The chatter rambled over farming's joys and woes, which might have been addressed in the past by calling a neighbor, consulting a reference book, or, now, bothering to conduct a Google search. Questions on the Active Farming discussion board ranged from "How can I identify the gender of rabbits?" to "What type of grain do I give my Holstein that I'm milking?" or "My dog drank hot chocolate!" But people enjoyed socializing and telling stories; they cherished this broad context. The quality of information being shared was secondary; I could see why scientists tended to dismiss the oral culture flourishing in this borderless frontier. An Active Farming woman who went by the pseudonym of "ShphrdGrl" listed her interests sweetly as "parenting, sheep, sewing, crocheting, gardening, and helping people."

Who wouldn't prefer the gentle ministrations of ShphrdGrl to my periodic lonely journeys through the 1,060 pitiless pages of the *Sheep Production Handbook*?

Meanwhile, at the Sheep Production Forum, serious shepherds analyzed how to hit the high price at livestock auctions, lamented American shepherds' difficulty in competing with New Zealand lamb imports, and discussed the merits of various feedstuffs. They paused in their colloquium to help a newbie who'd blundered into their midst—having missed the homey homesteaders on Active Farming by a country mile—as he thrillingly pondered what secretly was everyone's favorite topic: the choice of breeds.

The shepherds rushed in with advice about considering ultimate markets; they uttered warnings to avoid show sheep. The newbie understood that his role was to be grateful and repeatedly to admit his cluelessness. At last, however, he did what he wanted, informing them that he'd picked a breed because he liked its *color*, and he'd bought his sheep from what amounted to an exotic-animal menagerie. Although he'd wasted their time and utterly ignored their advice, the shepherds bore him no malice. Surely they'd soon be advising him on how to cure foot rot.

With the growth of the Internet and my desired niche as a breeder, I wanted to create a website that would showcase our farm and my efforts. I was expending time and money in acquiring superior rams and evaluating ewes and lambs. I needed to find more buyers like John Stenger who wanted what I was raising.

Our Web page's banner said Mossy Dell Farm. Although we'd sold it, the name was listed all over the Internet, in print directories, and with the Katahdin sheep society. And Kathy wouldn't let me name the hilltop farm separately anyway. After the sale of Mossy Dell, when I'd begun trying out Mockingbird Hill, having noticed my favorite songbird enjoying the trees and shrubs we'd planted, Kathy just shook her head. I understood. We weren't farming virgins anymore; we'd suffered and bled. Given our history there, bestowing a dewy alias on Fred's hard-used ground did seem delusional. Even at age three Tom had pegged my driven romanticism, back in Bloomington, when he'd removed his thumb from his mouth at the dinner table one evening to look at me—I'd been behind the house all day, slaughtering broilers—and make his first joke: "You a funny chicken." I dropped the notion of resuming my dream under a new pennant.

At least I had the perfect logo for our virtual farm's digital green banner. Our Freckles. In the photo she looked up and seemed to regard the visitor with friendly interest. She wore a brass bell on a leather dog collar around her neck, and what looked like a smile upon her lips.

CHAPTER SIXTEEN

Freckles had HUGE twins, a 11.4 lb. ewe and a 12.6 lb. ram.
—Lambing Notebook

SO WHAT INDEED, MOM. I SEE YOUR POINT NOW ABOUT SELLING Mossy Dell. Not such a big deal. There are endless dreams, so pick another or give an old one a new twist. That's what Dad always did, forever reinvent himself, a jazzed beginner chasing a glorious new dream.

And yet for me, Mossy Dell's loss—for that's how I couldn't help but see it, as a loss—was a sea change, another *before* and *after*. Like when Dad sold Stage Road Ranch and overnight we found ourselves living in a Florida beach town freshly scraped from palmetto thickets.

Kathy and I would never again own such a magical place; it just couldn't happen twice. Besides, we wouldn't try to make it happen. That chapter was over, already fading into our deep past. And our children were now poised to leap into their own futures, all in the eight years since we'd moved to Athens and I'd bought ewe lambs to graze.

Privately I wondered why losing Mossy Dell didn't bother me more. My whole life, I believed, that farm had been waiting for me, its perfect owner. Someone who saw its beauty, sure, yet who also understood how that beauty was entwined with its life as a working farm. As mystical and sentimental as that notion of inevitability now sounds to me, it still feels true, in the same way that Kathy and I felt, like any lovers, that we'd been moving toward each other across time and space for years before we met at Ohio State.

Something had changed in me, and I didn't know what. I didn't know why I wasn't grieving for the farm itself—whether my acceptance that Mossy Dell had left our lives was resignation or some kind of growth, however unbidden.

After we sold Mossy Dell, I read a letter one day in the *Athens News* from a newcomer who was incensed that his township trustees wouldn't force his neighbor to pay for half of replacing a brittle forty-six-year-old line fence. He threatened to withhold his taxes and mocked the officials, quoting one as saying, "Them laws can change from one day to the next."

Actually, Ohio's fencing laws hadn't changed in one hundred years. That legacy from a time when everyone kept the countryside and tended horses, cattle, pigs, chickens, and sheep was an anachronism that underscored how much farming and society had changed. As Ernie and Fred's feud over fences on our hilltop showed, only those who wanted a bitter fight now pressed this point. Fewer people kept livestock, and fences were overgrown with brush or had been ripped out to enlarge fields for crops. But people are territorial—at least I was—and need the security of clear boundaries.

So I understood the letter writer asserting his ownership. Yet now I could also imagine his neighbor living with the legacy of ancestors who'd tried to make money at farming and only got poorer, got hurt, got tired—got away. The complainer was technically right, but a fight based on technicalities is just manna for lawyers. And his snit revealed his own unneighborly ways. He reminded me of myself when we'd arrived from Indiana. Of my frustrations with life in a backwater, of my anger with locals' lackadaisical ways. I couldn't relate any longer to that Hoosier flatlander. What, I wondered, had he been so angry about?

Now, when I encountered a native blithely resisting an outsider's angst, I savored his lesson in Appalachian Zen. And the people no longer seemed completely ungracious to me. Reserved, maybe. I'd finally noticed that the men who seemed to stare rudely in Albany's feed store, where I bought grain for our poultry, were just looking to see if they knew me.

In summer 2006, two years after Mossy Dell's sale and ten years after our arrival in southeastern Ohio, I was serving as a trustee of Katahdin Hair Sheep International, was chairman of the Breed Improvement Committee and of the Shepherd Education Committee, and was a contributor to sustainable farming journals. My sheep society had registered its 50,000th Katahdin, which was becoming America's most popular sheep breed. Locally, I was on a panel studying ways to foster our region's economic development while preserving its prickly Appalachian soul. I was proud of my work and offices, a measure of how far I'd come and a challenge to how far I had to go.

My tractor's hour meter had logged 680 hours. That's paltry for a diesel engine—had I even needed a tractor?—and then I thought, *That's seventeen forty-hour weeks in the tractor seat.* Mowing pastures, pounding fence posts, towing my trailer loaded with hay or manure.

Sheep sales would hit $11,000 in the fall, decent for our intensively managed acreage with only fifty ewes. Later, totaling my expenses for the IRS, I was shocked, as always, by my costs. Yet again, I'd barely broken even. One of 60,000 sheep producers in America, I ran an operation similar to 94 percent of their farms in that it had fewer than one hundred ewes. Not big enough to cover costs. Ohio State issued a report that year on what it would take to make a living raising sheep: 1,778 ewes grossing $300,000, netting $78,000, and leaving $50,000 for family living. By then I'd never even fantasize about expanding to that scale, of buying or renting a larger farm, of again getting ambitious.

Trying to make farming profitable had united with aging to narrow my focus and make me cautious about big plans. I was an unapologetic specialist, a part-time grass farmer with one intensively managed species, one income-producing enterprise. I rarely imagined and never seriously considered starting anything that required new bodily energy or infrastructure. Adding cattle and range chickens to a sheep flock on pasture makes economic and ecological sense. But my new mantra was *No new chores.*

Farming's deep appeal remained, though, for me and my customers.

They were likely to be well into middle age and physically limited. I'd sold starter flocks to people who could barely walk. I marveled at these fellow dreamers yearning for vine and fig tree, for the clucking broody hen scratching at the barnyard dust, for the lamb drowsing on the sunny hillside. They came to me with their feet broken down from working at jobs on concrete floors, with knees grating bone-on-bone from age and obesity, with bodies twisted and in pain from automobile crashes.

I was, of course, a physically unfit agrarian myself, and increasingly overwhelmed by basic tasks. Daily chores might be compressed into only an hour if time was short, but time was always short. An hour's worth of time and energy might be all I had for the sheep. That August, after Tom joined Claire at Northwestern University—both my young helpers gone—some days the chickens seemed too much. I was forty-three when I'd brought our first lambs to Mossy Dell. I was late to begin serious farming and had less time than I knew before my body told me so. "You're only as young as your spine," my new yoga teacher said, and I'd worn out my lower back in fifteen years of farming in Indiana and Ohio.

Last year, late during our seventh lambing, the base of my spine felt jammed and then molten. After catching lambs, I retreated to the couch. Having taken time off work for lambing and farm projects, I lay around every afternoon in pain and drugged, trying to read Tom's favorite novel, *Crime and Punishment*. After two weeks of this, my right leg stopped working, and I dragged the limb as I did chores. In June, a surgeon drilled a hole in a lumbar vertebra and extracted disk tissue that was pressing on nerves. Before surgery I'd been depressed, but afterward viewed my aging body as I did my aging truck: periodic component failure was normal. I regained the use of the leg, though it was crampy, my ankle was weak, and the outer edge of my foot was still numb.

During my recuperation Kathy and I visited Sam in his trim gray house across the road. He hadn't been able to work all that season. While Erma told Kathy in the kitchen what they'd learned about his cancer—which still wasn't much—Sam told me, in their airless, mothball-smelling living room, his voice going high and breaking, "I'm going to beat this." He didn't sound like he believed it—he was saying what he thought he was supposed to say—and I felt somber, and guilty in my doubt. *Sam!* He'd always been so hardy, and his cheeks were still so pink, that it was hard to accept his sickness.

Yet death was all around. The previous fall Kathy and I had attended funeral services for Ernie's wife, Janet, whose long illness I'd never understood any better than I did Sam's. Ernie and Jim had asked me if Kathy would come to the funeral home, and seemed pleased when she appeared. I was proud that they knew Kathy as more than just another university big shot; she was also a neighbor who'd made Janet soup and bread.

Now Ernie had semi-retired. He was well into his seventies, and despite health problems was looking spry. One day, driving past his house, I saw him fetching bills from his swaybacked mailbox and stopped my truck in the road to talk. "Oh, I'm fine," he said. "I'm drinking my father's coal-mine tonic. Apple cider vinegar and baking soda."

Jim was still training horses and cutting my hair. My first farm mentor, Diana, was milking a goodly herd of Jerseys on the rich, rolling soil of northern Ohio and serving as president of the state's ecological farming association. I saw my shepherding role model, Mike Guthrie, around Athens sometimes; he'd sold his flock and land and was happy crafting cabinetry, gardening, and making beer. Glen Fletcher, unable to buy his tidy homestead from the Goss's offspring, who'd rented their farm to a cattleman, had made good on his dream of becoming a mountain man. He sent me postcards from Montana, where he'd bought wooded land near a stream and built a cabin. And Daniel, the haunted, driven, gifted excavator who'd rebuilt Mossy Dell's farmyard pond in our first lambing season, was still leading his men into battle against Appalachia's unstable terrain. He was past eighty.

Fred Paine went on, too. While Sam battled cancer, Fred, having apparently beaten the disease, could look forward to years of serial land abuse. Sometimes I'd see his faded black truck climb the hill in front of our house. I ran into him one day in the Albany post office. He loomed in my path, paunchy and unavoidable on the polished cement floor of the tiny lobby. Yet he was a footnote by then, ancient history, and I tried to greet his smug hound-dog's face as cordially as I could.

"Hey, Fred."

"Praise the Lord!" he sang out.

That tenth summer, as I piloted my creaking truck over the narrow, rutted back roads, my mind drifted to the period right after we'd moved to Athens. I'd started at the press, and in my spare time was doing some freelance work for *Scribner's American Lives* encyclopedia, writing profiles of famous authors. For the first time I'd found myself thinking about how, fourteen years before, I'd left one of Florida's best newspapers to pursue Kathy and had started over on a small-town Hoosier rag.

I hadn't ever regretted doing it, following my lover. Yet I became, in that act, "the trailing spouse," as they said in academia. Someone who lacked a true career but who worked anyway, for self-respect and supplemental income, once the woman's role. In the upheaval of our move, I'd asked myself for the first time, What kind of man would do that? What kind of man would throw away his career? Couldn't we have had a long-distance relationship until *she* got a job in Florida? That option hadn't occurred to me at Ohio State. I'd leapt for her without one doubt when I was alone and lonely. Kathy had redeemed the world. All the same, my choice didn't jibe with the swashbuckling men and women I was writing about.

Now, with a secondary but respectable career and a squared-away farm, I wondered why I was still so anxious. Who was this fretful guy, as timid as a sheep? One day driving into work, I noticed I'd eaten my fingernails yet again, probably over my expanded duties at the press. But farming's difficulty still upset me, too. And I couldn't understand why I'd panicked five years before and lifted Red into my truck.

"Let me tell you a secret," Joel Salatin writes in *Family Friendly Farming*. "We all have baggage. The question is 'What am I going to do with my baggage?' I think every person's goal needs to be to work through it and throw out the unnecessary stuff." I had trouble letting go of pain, anger, guilt—I'd begun to see that. But I also saw how my anxious nature itself created, in effect, daily baggage. Of course I'd noticed that Kathy got up every morning and marched out to take on the world, while I struggled with myself to get through an ordinary day. *Is everything okay? How am I feeling? Is anybody mad at me?*

Temperament is usually what people like Joel really mean when they talk about "character," but it has nothing to do with character. It defines, instead, the nature of one's suffering: impatient and then angry, or anxious and then fearful. One's response to the world can be managed but not jettisoned, I concluded. Animal trainers and good stockmen understand this, know they must get an animal when it's young to have a chance of modifying reactions that are too aggressive or too shy. Our Jack was my own best example. The little terrier had come to us as a raging Type A—he'd growl and snap if we displeased him—and in my alarm I'd dominated him physically and emotionally. I hadn't abused him, and he was still feisty, a game little hunter, but Jack became sweet and submissive with us.

People are turned, as well, toward one end of the spectrum by experiences early in life with siblings and parents and situations. It takes all kinds to make the world go round, even those on the anxious and self-critical end of the bell curve, but I'd wanted to be like Louis Bromfield, a force of nature acting in his own epic agrarian drama. Or like Joel Salatin, a questing pioneer at the helm of his impossibly busy and improbably diverse farm down in Virginia. Or, always, like Dad.

Farming had toughened me—I liked to think I'd become a hardened Darwinian practitioner, no quivering romantic. But such change is relative. I had to admit I was overly sensitive, that I overreacted all the time. Anxiously lifting Red was the defining instance. My dreaming boyhood had fueled my farming adventure but couldn't prepare me emotionally for its challenges. Even if we'd kept Mossy Dell and I'd grazed many more sheep, my farming still would've been part-time—all I was comfortable with.

So what had Mossy Dell meant? Maybe that I valued experiencing a lovely landscape as much as I did farming it. I decided, during my travels to and from Athens that summer, that losing Mossy Dell didn't shatter me because I'd finally stopped chasing a dream for which I was temperamentally unsuited. I wasn't a man of action, but someone obsessed with making sense of life, which felt, moment to moment, like swimming in a river of risk and beauty. My sensitivity was what Kathy said had attracted her at Ohio State; now she thrived in her career partly because of my belief in her, and she relied on my insights. We wouldn't have met if I hadn't left the *Orlando*

Sentinel for that fellowship, drawn to Ohio by Louis Bromfield's mystical dream of farming. And while a tougher man might have held onto Mossy Dell, not let it slip through his fingers, I doubted a hard-fisted man could've taken that old farm into his soul. Certainly such an attachment was something that Dad, forever passing through, couldn't afford.

I'd lost a fantasy about myself, one vision of who I was or might become. That wasn't so bad, for with losses came gains, my new self-knowledge. Anyway, loss was inevitable. As Dad would've said, it goes with the territory. He carried loss through the Michigan snow to the California desert, to coral atolls in the glittering Pacific, to the firebombed ruins of Tokyo, to south Georgia's red clay flatland, and finally to Florida's unruly interior.

I was proud that I'd acted, at last, on my dream. I'd overcome my fears and followed my passion, did what I thought I wanted to do. To have remained a dreamer would have been to stay in some sense a child, that wounded boy in his bedroom in Satellite Beach, daydreaming about growing up a cowboy on the California ranch, or pining for the gracious Georgia farm, grown mythic from my earliest memories and Mom's stories.

I couldn't regret anything I'd done, not even my worst mistakes. I'd become a farmer. I'd found my farming niche. I had a good woman and two great kids. And I knew myself better. I could live with that, accept that— accept myself—in relief and with gratitude.

I walked the gravel driveway to the barn, and couldn't believe it was already October. The leaves had changed, though, and big wolf spiders again startled us by occupying our house for the winter. They crept, regarded us like crabs in a sideways dodge, made a chitinous crunch underfoot. *Where did September go?*

The morning was chilly and clear, the sky a brisk cloudless blue, the hilltop's everlasting breeze just awakening. As I stepped off the porch, a pair of mallards leapt from the gravel in front of the barn, the hen with a loud quack; they flared overhead, the white undersides of their wings flashing, and rocketed toward Lake Snowden. I was wearing a gray fleece work shirt and a green knit cap, dressed for my mission of clearing the barn's central aisle so that tonight I could tow in a trailer loaded with bales of hay and stack them inside, safe from the low November sky I knew was coming.

For years I'd kept the barn's center clear so trucks could enter the front door and leave out the rear. But last spring I converted the aisle into a temporary sheep pen. Now it was time to reclaim the area. I dragged apart my makeshift corral and leaned sections against the walls. Suddenly I thought of Sam. I pictured my hired hand bustling across the aisle, moving fast on his bowed legs, his ruddy face shining in the barn's gloom. We'd always set up the barn for a new season together.

Sam's death is what happened to September.

He died three months before his seventieth birthday. One reason I moved sheep into the travel lane, I realized, was that Sam fell ill: I needed a clean pen because, without his nagging, the ram's stall and the lamb pen on the north side hadn't been mucked out in three years. There was a tight hump of melded manure and hay two feet thick over there, and to avoid deepening the pack I started the new enclosure. I hadn't thought of Sam since the funeral three weeks before; now, working in the barn without him for the first time, I felt lonely.

The chickens noticed me stirring the dirt, and the birds came running. This free-range component of our flock was getting out of control. Tom had been responsible for chicken chores, and now, surrounded by chickens, I realized he'd been throwing hens out of the coop for years. He probably was mimicking my passion for selective breeding by adding hens to the consorts I'd approved for a few bantam roosters that decorated the barnyard. The result was a population explosion. Chicks—young for this late in the year—were running underfoot. Hens hurried with their broods into the barn to scratch in fresh spots I'd uncovered. One, her plumage spangled white and red, flew with her equally flashy chicks, the size of quail, onto a metal gate I'd leaned against the manger. The birds perched there to survey my activity, but they upset the gate and it crashed. The hen shrieked and fled with her offspring. I lifted the gate and a chick popped from beneath and scurried away with one wing drooping.

Heaps of old hay moldered atop pallets in a corner I needed to clear.

I realized I would uncover rats beneath the mess and should fetch Jack. Rodents had infested the farm when Fred stored corn here—all the outbuildings, littered with droppings, had stunk of rat urine—and they'd crept back with a vengeance. My pallet floor was a rat paradise. Without my help, Jack couldn't reach them in their refuge in the slats. The little monsters had ready access to food: laying pellets in the henhouse; grain in the trough in the aisle for the loose chickens, guineas, and mallards; and in winter, cracked corn in the north bay for held-over slaughter lambs.

The rodents were also benefiting from the manure pack, where they'd nested in voids and done some tunneling. I'd been forced to accept that big, fat Norway rats inhabited the barn—that real farms had rats. I never told Sam of this worsening problem, which would have horrified him.

Jack helped me keep pressure on the vermin. And they gave him a purpose in life: he was happiest chasing rats. When he was unsupervised, I kept him tied on our porch because he now dug out of his and Doty's kennel in the open-fronted shed and left gray-muzzled Doty behind. I panicked whenever he escaped, afraid he'd burrow alone deep into a groundhog's den, where the giant rodent might eviscerate him.

As I walked toward the house to get him, I glimpsed the year's surviving guinea-fowl hatch running across the lawn. I tried to count the keets but failed. As the months passed, Kathy and I watched the broods dwindle. Having evolved on the dusty trampled plains of Africa, guineas, unlike chicken hens, made no allowance for morning dew, and dragged their young through the wet grass. The downy keets, toddling on orange toothpick legs, became sodden and chilled, got left behind and died, a snack the size of a dandelion's puffball for other creatures. Even in Africa, the guineas' purpose must've been to supply these protein morsels; otherwise, screaming guineas would overrun the world. We'd been able to keep the flock going for almost a decade because the birds roosted in the barn's rafters, sixteen feet up, safe from marauding raccoons.

Jack knew what was up the instant I freed him from his tether, and he raced ahead. "Get those rats, Jack!" I yelled. He was a white streak, eighteen pounds of muscle and bone quartering through the barn. His bowed front legs, broad chest, and bulging shoulders gave him a Mighty Mouse look. He shoved his nostrils into holes, took a whiff, bent to dig. His butt stuck up and his abbreviated tail wagged furiously.

I lifted a pallet for him and rats scattered. Jack bit one, which squealed; he jerked his head and it went limp. I searched frantically for where I'd propped the pitchfork. Big rats, trophy-sized, were getting away. Jack had his nose in a hole and rats were running past him. "Jack!" I shouted and came running with my weapon. He was intent on a rat that was out of reach in the slatted flooring. I did the rat dance: holding the pitchfork at waist level, I staggered over wads of hay and across uneven pallets, jabbing at rats. One ran blindly toward me but then dodged. I thrust at another, stabbed the ground. Rats skittered past. I poked at one running along the base of the wall and jammed my prongs instead into wood, and then wheeled in a new direction with my spear. I must've looked like a drunken man with palsy, reeling in a corner and shaking a pitchfork's tines at phantoms.

Jack killed about four rats. At least a dozen escaped into the pallets. Although I hadn't scored once, I was irked by his inefficiency. Yet he surely considered me inept. Each of us had his limitations: I could see the rats but couldn't move nimbly enough to kill them; Jack was fast but kept his nose to the ground when he should've used his eyes. He never paused above the fray to see the big picture. He wasn't an ambush predator, like a cat, but went directly after prey, his front feet a blur as he dug. He used his mouth, too, impatiently biting clods, roots, rocks, logs. He'd broken three teeth, and ripped out two by the roots.

Now he was barking in frustration. *Come out so I can kill you!* His nose, shoved constantly into trouble, was bleeding. It never healed, always crusty with scabs or the tender pink of fresh scar tissue. Tonight, collapsed on the couch, he'd give me and Kathy dirty looks, as if to say, *I'm tired after working so hard, but you're both fresh as daisies.*

I was hot, sweating, and walked to the house to change shirts. Outside the barn it was becoming a perfect blue-sky fall afternoon. The two hickories across from our house had changed colors. In summer they were a pair, the tips of their branches touching, identical dark green ovals on our hill-top. Now one tree was golden brown and its sister was yellow with green highlights.

The hill fell away beneath the trees and met our wet northern pasture. This bright green slice contrasted with Ernie's overgrown hill that rose beyond. Relinquished to swaths of yellow goldenrod, its steeper slopes bris-tling with thickets of red sumac and saplings, now as tattered as battle flags, the round hill looked beautiful against the fresh grass of our low rectangu-lar paddock. Ernie's land was in limbo between farm and forest. Below his wild rolling fields and our manicured hilltop, Ridge Road snaked under autumnal foliage and curled past his abandoned silo. I loved this northern view, and felt more like a painter than a farmer as I gazed upon it. I always wondered whether it was the contrasts or the decay of human intent I found picturesque.

Every fall, when the air hazed and the far hills turned purple, I paused and looked down from our hilltop at the green grass, brown and yellow weeds, scarlet slashes of sumac, pewter silo, and faded-gray country lane. The curving road was a never-ending story: *What's next?* Our first summer on the hilltop I saw a big red-tailed hawk flying across this tableau, carrying a writhing black snake in its talons. The flesh-eating bird and the ground-hugging serpent, flown aloft to its death, materialized like an Aztec icon in the sky. That image haunted this common stretch, kept me searching a derelict landscape as if an answer—to what?—might dart like a rabbit from Ernie's weeds.

I returned to the barn wearing a fresh shirt, a thin denim jacket, and gloves to protect my hands while lifting pallets. It was after ten o'clock, and I feared I might tire before finishing today. But I'd already lined up a friend to help me fetch hay. I heard a sharp bark, Jack still working at the far end of the barn.

He looked at me expectantly, and the slow wag of his tail signaled he was waiting for help. I lifted a pallet, but he wasn't interested. By looking at me and then at the wall, he indicated he wanted me to remove a sheet of tin that Fred had put there as a baseboard. I leaned over and saw that two pieces were wedged together, and the vee formed where their tops leaned away from each other was crawling with rats. Jack had treed countless rodents in the steel valley. I pulled it apart and rats scattered. Jack killed two while I yelled, but fat ones disappeared in the melee; the big ones were slower but must've been smarter, because they escaped more often. Jack seemed to take the different sizes as they came. Apparently he enjoyed the feel of any size rat body against his gums. He should have favored the junior ones, because they were too small to curl upward and bite his nose before they died. Jack cried when a rat bit his nose, but never turned loose.

Now I laid down pallets where Jack, Kathy, Tom, and I had a good rat hunt last spring. Acutely aware I was creating a new rat refuge, I left Jack a space between the skids and the wall, so at least he could circle the rodents' metropolis and bark his angry bark in their ears. As I worked, I yelled encouragements as he buzzed past.

"Get it, Jack!" "Is there a rat there?"

Finally, his sides heaving, his tongue hanging, his snout bleeding, Jack headed for the black rubber pan under the barn's water hydrant. He lapped water and then plopped down with a splash, cooling his belly. Curled in the shallow pan, he panted for breath. Wet, he was going to get filthy in the dust and manure. I started walking to the house and called him—"C'mon, buddy. Let's eat. It's lunchtime!"—but he didn't follow.

He stood dripping in the pan's milky water, his brow furrowed, his olive-brown eyes staring as if my face held the answer to Life, his unwelcome realization dawning: *My farmer's ending the hunt.*

After lunch I carried an issue of the *Stockman Grassfarmer* into the living room and sank into the leather La-Z-Boy recliner Kathy got me after I was named the press's marketing manager. The magazine's lead success narrative concerned a dairyman who was failing until he started "adding value" to his milk by turning it into cheese. His wife and daughters sold it from the farm's homey roadside market. The story didn't grab me; I was never going to make cheese—or anything like it. I thrust my throne into full recline, closed my eyes. My sinuses were clogged, my knees were sore, my lower back ached, my neck was stiff, and a muscle was clenched below my right shoulder blade.

Startling awake, I lowered the La-Z-Boy with a thunk, struggled to my feet, and shoved off. As I drifted to the barn I thought, Things have a way of getting done. The sorriest rural hovel was kept from complete disorder, saved from the encroaching forest, by human activity, both aimless and fitfully purposeful. Shrubs from K-Mart engulfed the front stoop, but eventually someone got pissed and hacked yew branches off the concrete steps with a dull machete, or chopped at the offending greenery with a rusty sling blade. People kicked aside aluminum trays the hound had licked clean, and paper sacks that had held chicken feed. The dog escaped its chain often enough to keep possums from nesting in the debris under the porch. Someone sprayed half a can of Raid on the hornets nesting under the eaves.

With Sam, as pertinacious as Jack after a rat, tasks got done. Even without him, I saw that the gravel driveway had once again reached the end of the season without complete takeover by weeds, thanks to drought, vehicles, and a little Roundup from my backpack sprayer. It was obvious I shouldn't have been stressed out by the disappearing driveway. Some things became invisible the more I saw them, and some didn't. Sooner or later, the lawn would get mowed. The driveway, its borders blurred by vegetation, would stay reasonably navigable. I knew that this hope that the imperatives would be accomplished—with or without an agenda, or in the collapse of successive to-do lists—was an attempt to comfort myself. I had so much to do with winter coming on.

The sprayer itself was out of commission. I'd ordered a repair kit in late summer and then lost track of it. Surely I would've placed it beside the broken device, but it wasn't there when I went into the garage and stared at the sprayer. Later I discovered the unopened cardboard shipping box on top of my dresser, invisible where I'd placed it. The fight for the driveway was over for the year, after all. I knew that, even as I tried to lash myself to do the right thing and fix the apparatus before I needed it again. Sam once borrowed the sprayer, which held five heavy gallons, and helped wear it out by spraying his lawn for several days. I couldn't imagine how many gallons he ran through it. A man who never complained of aches and pains, Sam admitted he was sore after that marathon.

Back in the barn, with my temporary corral removed, the sagging gate across the barn's rear doorway confronted me. I must've rehung that gate three times, but it always drooped. I wanted to raise it, so that I could swing it open easily and drive my truck and trailer right through. This was an excuse to get to something that had been bugging me, one of those small improvements that made me feel unreasonably good while being ancillary to the actual job at hand. I suspected that such optional tasks were always my real motivation, my favoring dessert over the main course. I justified the fun job as my reward for the tiring and ephemeral cleanup—before I knew it, it'd be spring and I'd be tidying up again. I decided to use a trick Sam showed me and defeat gravity once and for all. This took tools, also a steel hinge pin, and I plodded to the garage.

I loaded our garden cart and dragged it back, heavy-footed. Six pigeons, streaking through the blue air for the barn, swung wide when they saw me coming. On the barn's naked steel roof, orange islands of rust were spreading cancerously. The barn would look great with a new green steel roof, and with its pitted corrugated tin siding, sprayed a dull red like the sheds, replaced with local poplar. This gentleman-farm makeover was an old fantasy I hadn't gotten to. And as I watched the pigeons arrow away, I knew I never would.

I knelt stiffly and drilled a hole into the wooden doorframe for the pin, which instead of being used as a hinge would serve as a rest for the opposite end of the gate. If I placed the peg just right, I'd be able to lift the gate's foot-stub of hollow pipe over it; that way the peg also would act as a strong, simple latch, so the gate wouldn't need to be chained. How proud Sam was the day he insisted we use this embellishment for a sixteen-foot gate into the south pasture; it had never sagged.

Sam had seen this idea at a friend's farm. As I worked, it occurred to me that I hadn't thanked him sufficiently for the innovation. At the time, he probably seemed too pleased with himself and annoyed me. Why, I wondered, did I withhold from Sam my full appreciation? Then, still on my knees, I thought, *This is just guilt. How can we ever value each other enough in the face of death?* I *had* fussed over him, I recalled, even though I'd been unsure about his latest enthusiasm.

When he got sick, Erma said she thought his rare cancer was triggered by the noxious chemicals he'd dispersed in his house and yard. I found this believable, given my own sensitivity; I broke out in hives just from catching a scent of herbicides on the breeze, and limited my own poisoning to sparing driveway use of comparatively innocuous Roundup. But Sam was hardier than I. He was almost never sick—he was a big baby when he did catch a cold—and came from long-lived stock. His mother had survived into her nineties, and his father spent a long lifetime farming and running the feed mill in Athens.

Based on his genetics alone, I'd been sure Sam would outlive me, celebrating his hundredth birthday in a nursing home, gumming his birthday cake, his mind long gone. I would picture this: Sam's eyes alight, a shiny blue cone strapped atop his flossy white hair, white frosting stuck to his severe mustache.

At Sam's funeral service in the Albany Baptist church—the village's largest structure, a rambling compound of brick and steel—a retired teacher had told me that Sam talked often during their card games about our projects. Maybe Sam spoke of my work as a grazier, controlling the wandering of sheep and harvesting grass efficiently, although I couldn't recall ever telling him the theories behind our efforts.

The fine print didn't matter to Sam. He liked to work. His oral culture and subsistence heritage valued the traditional rhythms of life that arose from tending plants and animals. Sam had learned that farming didn't pay, but the lost world that had produced him had known poverty, even hunger, and such hardships affirmed the importance of land ties. Maybe he never asked about the details that obsessed me because he understood what I was doing better than I did. The desire to farm didn't puzzle him.

We gathered for his rites in the sanctuary, a vast space that reminded me of a basketball gymnasium—an effect magnified by the fleet of yellow school buses parked outside that carried Baptist youth hither and yon. I stood in front of the pulpit in my blue wool suit that I wore to Chicago and New York on book-marketing trips. Behind me on the wall were huge white screens, now blank, for projections of Christian rock lyrics. The mourners faced me in rows of chairs on plush blue carpeting.

"Sam was a Christian," I said. "As much as anyone can truly be such a difficult thing, Sam was. He was a man of faith. He kept the Sabbath, and he tried to love his neighbor. He tried to avoid trespassing upon others, and he tried to forgive trespasses. He was naturally a good man, and his faith made him a better man."

I didn't mention that, unlike me, he was always cheerful. Or that I was grateful he didn't seem to judge me. I cited loving one's neighbors and forgiving trespasses because they were my favorite biblical precepts; the fact that they made sense but were nearly impossible seemed to prove their wisdom. Of course Sam held grudges and condemned lazy neighbors. And I'd always lamented his observance of the Sabbath, a whole day lost to farm projects. Yet it had resulted in my family seeing me more; it had put our human toil into perspective, as intended.

"Sam was an inspiration to me," I ended. "He was my friend, and I'll miss him."

I realized, as I spoke, that these things were true.

Sam's minister, a handsome man with a golfer's tan and a pompadour of bronze hair, shot a downward grimace at me as he mounted the pulpit. "We've just heard about *acts*," he said. "Good *deeds*. But no man can enter the kingdom of heaven unless he has accepted Jesus Christ as his personal savior. A lot of good people aren't going to heaven. Sam and I discussed this. He didn't give me the details, but I could tell he had been saved."

I should have known better than to poach on the preacher's turf by making scripture mine. Especially after my boyhood in the Southern Baptist church, where Genesis was read literally, where someone was saved every service, where the Plexiglas baptismal tank was kept sloshing. Even so, his rebuke stung. And yet, knowing Sam, I felt vindicated. Sam had let the man think what he wanted, but Sam was too shy to publicly testify or to undergo the trauma of getting dunked. He surely was fudging, figuring God would look down and give him credit for occupying his seat every Sunday beside Erma.

After the service, a man in a stiff maroon sport coat made his way to me. It was Ed McNabb, Fred's hunting buddy, the farmer from the other side of Lake Snowden who'd clashed with Ernie on our hilltop five years ago, each pleading his case in turn with my visiting mother. He shook my hand and thanked me. His wife, whom I'd never met, looked at me hard and said, "You did a *good* job."

That was acceptance, I understood. And there was the obvious: that I'd finally known a local man well enough to speak at his funeral. I would always be an outsider, but I'd been seen. And I understood that part of the genius of the place passed out of memory with men like Sam. All the same, I knew I'd failed to see what Sam meant to me when he was alive.

When he used to show up on my porch for work, as I went to answer the door I'd see him fidgeting. He'd be early and holding a wax-paper packet of cornmeal mush he'd made for me, and foil-wrapped treats for Doty and Jack. With dew still on the grass, he'd have stretched over his boots low-cut black rubber galoshes, surely the real secret of his immortal leather.

I would make him visit with me, dragging out the day's start. But he'd want to get busy and soon would utter my favorite of his expressions, his Appalachian Zen retort to my demons and a reminder that perfection lies beyond us.

"Let's do something even if it is wrong."

Then we'd step into the sunlight together, happier than we knew.

EPILOGUE

To him who in the love of Nature holds
Communion with her visible forms, she speaks
A various language
    —inscription on Louis Bromfield's gravestone, Malabar Farm,
        from "Thanatopsis" by William Cullen Bryant

THIS APRIL MORNING IS SO MILD, THE SPRING SO TENDERLY ADVANC-ing, that I'm surprised the forested hills remain bare. The naked trees, rising above lush pastures and weathered crop fields, are the dry mousy color of deer. Over the gray-brown domes of the woods there's a golden-green haze: budding leaves. In field borders and woodland edges, brush is in full leaf; the multiflora roses glow lime green. In lawns, growing fast now and clumpy, whips of forsythia arch in ecstatic yellow sprays. The breeze smells of cut grass and gasoline: someone has mowed. The airy white blossoms of pear trees float in yards. Over at Mossy Dell, Mabel Vaught's saucer magnolia will be covered again with pink flowers.

I find Freckles collapsed exactly one week before her due date, as wide as she is tall. I summon the veterinarian, who thinks it's milk fever, a metabolic imbalance from her fetuses' demand for calcium. We get her into the barn, my tractor ferrying her comatose body on a piece of chipboard I've strapped atop two steel prongs I use to move round bales. After the vet shoots her full of calcium and energy boosters, she rallies, and is able to sit upright in the stall.

The next day, sometime between my rounds, Freckles rolls onto her back. She gets cast—unable to stand, a capsized turtle—and stuck on her back, her own weight suffocates her. In all her years with us, she's raised only six daughters: four sold before I knew what I was doing, one strangled in a fence during her first lambing, and one retained, a yearling.

Old Mama has pumped more daughters into the flock, though her lumpy udder is mostly scar tissue from mastitis. Now it's a spring ritual to run her into the barn and train her lambs to drink artificial milk from a white plastic bucket. She gives them emotional support. She's brusque and appraising toward me, all business, but she understands I'm feeding her lambs under her nose. Skeptically she supervises my efforts; doubtfully she watches her lambs suck red rubber nipples. I've grown certain she's Freckles's mother. Aside from their physical resemblance and their overlapping time-lines in the Goss flock, they share that astonishing maternal ability.

The flock still sports some huge Goss ewes, to satisfy customers who want only size, but gradually it's dominated by Old Mama's compact daughters and granddaughters. And one of them, one of Freckles's sturdy half-sisters, has delivered a son, sired by Kansas, that's the first homegrown ram I feel confident enough to breed our flock to.

Only I won't.

Kathy is on the job market. Caught in a provost's classic dilemma—between an embattled president and an angry faculty—she's started responding to queries from executive search firms.

So this September of 2007, getting ready for our eventual move, I begin to sell down the flock.

Serious breeders come first, a woman from Pennsylvania and another from northern Ohio. They sort through the ewes as I rattle off data from my clipboard. One buys Freckles's daughter. Next a man tows a long red trailer up our hill and loads many more ewes and a ram. He represents a consortium of breeders from Maine, home state of the Katahdin breed.

Hoping to raise a few more rams for regular customers, I keep nine ewes, including Old Mama; I can't sell, or even give away, a ewe without a working udder, no matter how valuable her genetics. To willingly create bottle babies? Few shepherds are that crazy. Several years ago, I'd have sent her unsentimentally to market. Now I know her breeding value, the great virtue she passes on to her daughters, that fierce love for their offspring. And anyway, I'm fond of her, and it's getting harder for me to kill.

The night the man from Maine drives down the hill with most of our ewes, I have a nightmare about Dad. Talking to me about farming, he says something critical about the sheep: "They're a lot of trouble for what they're worth." I roar back, "I did better than you did at farming! The sheep paid their way and then some." He looks stricken; tears wet his cheeks. He hadn't meant to be disparaging. I'm shaken by what he'd said—by what I've imagined—and by my response. My anger shocks me, and more so my rivalry. I've never seen that while I tried to emulate him, I also tried to outdo him.

There was noisy bulldozing going on in the dream, massive destruction of some kind around us. I feared something had been bulldozed by mistake, but I couldn't see clearly for the dust and rubble, everything knocked flat or shoved into piles.

Our tenth and last lamb crop is due in one week. Until this Sunday I've rebuffed Kathy's recent desire to attend the country church near our house. On the cusp of leaving, we hunger for community here. Maybe we feel, at last, that we've earned our place.

"I'll go, but I don't want to start and then leave," I protest as we climb into our van. "We have to stay, if only as a spiritual discipline. Until you get a new job."

"It's Presbyterian, Richard."

"Who knows what Presbyterian means out here." I'm wary after my experience with Sam's minister.

But Kathy's got a good feeling about the church. And while we'll surely never have another gifted preacher like ours back in Indiana, maybe that doesn't matter. Anyway, we're ashamed of our sermon-shopping through four Athens churches. We admire friends who've stuck in their pews, enduring shifting ministerial talent.

Now Kathy elbows me in the ribs. Her red hymnal's open to the first song, "Morning Has Broken." I'd learned its words back in Indiana so I could sing it to Tom every night as a bedtime lullaby. I look around the sanctuary. The crown molding is the same local style that was crafted for our house from Mossy Dell's fallen oak. Behind the pulpit there's an ornate lighted windowbox in which a sandy-haired Jesus, his soulful eyes gazing gently upward, serenely abides.

We've never before set foot in Alexander Presbyterian Church, founded in 1832, though for years we've admired it. With its Gothic stained-glass windows, pristine white clapboards, modest steeple, and red tin roof, the church is a landmark beside the Appalachian Highway. We visited its cemetery once when Tom was in elementary school, because for a class geology project he was identifying different kinds of stone. I'd heard that Mossy Dell's previous owners were buried here, so I searched for the Vaughts amidst acres of markers.

Kenneth's and Mabel's modest tan granite slabs stood on either side of their daughter's; Betty's was three times as large. We saw from the dates that she'd died two weeks before her sixteenth birthday, in 1943, and I remembered Jim's story in the barbershop about a softball hitting her in the stomach. Kenneth had lived on for forty-four more years. Mabel, only eighteen when she'd had Betty, I noted, died at eighty-six, having survived her daughter by fifty-three years. In front of the family's plot loomed a large gray stone for twenty-four-year-old Delbert Vaught, Kenneth's younger brother, killed in France in 1944. A hard blow that must have been, the year after Betty's death. Standing in the wiry cemetery grass, I'd tried to imagine the region in wartime. Just another rural outpost, yet with stories going on: love affairs and heartaches, rumors and headlines, heat and flood, church suppers and Easter egg hunts.

Twelve years ago, in 1996, just after Mabel Vaught was laid to rest, we'd driven past on our way from Bloomington to Athens. We couldn't know that we were on our way to buy the Vaughts' beloved farm. We couldn't know that in a cemetery a stone's throw away, another little family's story had just ended.

Sitting in our pews today, back from Kathy's latest job interview, we're tired. Having settled into our empty-nest lives, everything in this spring of 2008 feels urgent and overly significant. The choir enters, elderly women robed in bright blue, joined by the minister. Pastor Bob is a large, mild, silver-haired older man with a benign round face. He wears eyeglasses and a black robe; his white stole is marked with the word Joy. After singing "Morning Has Broken," we shake hands with almost everyone in the church, forty-three souls, who erupt in talk and laughter as they circulate.

The scripture lesson is from the story of Paul's journey through the East; he'd wanted to start his own church, but instead was sent to proselytize in the sticks. "If it fits you, listen," Pastor Bob says. "Maybe I'm just talking to myself today." His sermon is "Handling Second Choices." He tells us that as a young man he'd been preparing to leave the four small Methodist churches he was serving in southern Indiana. He had a pregnant wife and a toddler. By uprooting for Duke University's seminary, he'd advance and reside closer to his parents and brother. One afternoon before leaving he got a blinding headache as he walked into his house. The pain was so intense, so otherworldly, that finally he prayed, promising he'd do anything—even not leave—if only the pain would pass. It vanished. So he'd stayed, and ended up in Ohio. A couple of years ago, he'd been eager for retirement from a Methodist church just north of Athens. Then he heard Alexander Presbyterian was struggling without a minister.

"This church seemed to need me," he says. "And I needed this church."

His boyhood dream, he confesses, was to become a baseball announcer. "A teacher said I had a good voice. I hope you all don't mind that you were my second choice. Sometimes second choices are the right choices. The only way you're ever going to get over being second choice is to find people who need you and help them."

I nudge Kathy's arm: *He's good.* And maybe, despite our regrets about leaving, we're needed somewhere else. Anyway, I think, wherever we go is where we're supposed to be. Not in the past, not in some better future. Lately I've been meditating with Athens's newest clergywoman, a Zen priest, her head shaved as bald as mine. "I *will* end my suffering," she told me. Her suffering, she said, is anger.

Forgive. Love. Create. These are the words I think when I meditate. My body now thickened and scarred, I feel more resilient, ready perhaps for an inner journey.

After my nightmare about Dad, I'd called my sister, distraught. Meg was closer to him than any of us except Mom. I told her of my shame from yelling at Dad. But instead of focusing on that, she went to the bulldozer. "It was clearing the way," Meg said. "Making room for something new. That dream was about rebirth."

In November I point out Old Mama to my final customer, a local man who last year bought a starter flock of thirty ewes and a ram—a friendly, swaggering chocolate guy I named Elvis, another of Old Mama's sons. As we load the farm's last lamb crop, I admire Old Mama's daughter we raised this season on the bucket. She'll make a beautiful ewe, just like Freckles—white with tan spots, black flecks across her nose, red hocks—and she already seems to have her mother's bemused attitude toward me.

I've sold over $30,000 worth of farm equipment and sheep, mostly breeding stock—finally a banner year, which includes a $329.50 payment from the federal government for doing something. For farming.

The next morning I find our last sheep dead in the barn lot. Old Mama was at least eleven and probably older. She'd been moving slowly yesterday, or so it seems in retrospect. I've always said I'd let her stay till she dies, and now she has. I'll avoid the silent barn, sparrows creeping in the dark rafters. I'll leave the last sheep on the farm where she lies, food for buzzards, the end of my farming career.

As Thoreau tartly commented in *Walden* at a time when most Americans were still growing food, "The farmer is endeavoring to solve the problem of a livelihood by a formula more complicated than the problem itself." As this implies, farming is yoked to humankind's fall from grace, to the loss of Eden—to laboring for our bread in the sweat of our faces—but farming also fueled the risen glory of human civilization. Yet it's fair to ask, with Thoreau, why anyone wants to farm. Especially in a post-agrarian, postindustrial age that has liberated affluent people from the perceived drudgery of farming.

America has rendered its judgment: in 2005 the nation became a net food importer for the first time. With a mainstream farm system based on industrial models and standards, information-economy America was importing increasing amounts of fruits and vegetables; it appeared likely that we'd soon begin buying most of the grains that underlie our culture.

And in 2006, Joel Salatin, whose farming in the Shenandoah Valley had become famous in the dozen years since I'd visited him to learn how to raise broilers, made an alarming observation in one of his articles: the number of people incarcerated in America's prisons surpassed the number of those growing its food.

Yet to my surprise, in the dozen years since we'd moved to Athens, the farming world had changed in another way, a positive one, with the return of a cyclical countermovement. Suddenly there was another herd of people, including prosperous young couples and retirees, affirming farming as a way of life. Even major agribusiness publications reflected this. Dad wouldn't have recognized his *Progressive Farmer*, which looked like a lifestyle magazine—not quite *Southern Living*, its sister publication, but getting there. (I had to laugh, though, when *Progressive Farmer* ran a clunky historical overview headlined "1910–1919: The Golden Age of Agriculture." That's it—*nine years*? It seems that was the first and perhaps last period when the buying power of farmers equaled that of the general population.) There was even a slick magazine, full of advertisements for compact tractors, unashamedly calling itself *Hobby Farmer*. With my anxious nature and my spine riddled with bad genetic information, where was that outlook when *I* needed it?

Many of my breeding-stock customers had this broader perspective from the beginning. They didn't aim to make money. They came to farming seeking aesthetic pleasure and solace from an angry world. And a word had arisen to honor food produced with less control but more craft: artisanal. The goal wasn't high production per acre, but food infused with love and time. Like art.

Can big mainstream farms produce boutique nourishment? Maybe, with great management, sophisticated technology, and cheap labor and fossil fuel. But size, as I'd learned in raising sheep, inevitably is hostile to something else. For the highest quality, nothing beats small, slow, and inefficient.

Many people want to enjoy animals and grow some of their own food. Those are elemental urges embedded in *Homo sapiens*'s genetic makeup. We may be a fallen species, forever barred from returning to an effortless cornucopia, and yet so many of us hunger to create our own little Edens. Are we really post-agrarian? In our livelihoods, certainly. Not in our deepest desires.

Of course I shouldn't have reached for Fred's cornfield. I see that now; surely I knew it then. Or, having gotten his home place with it, we should've sold his trashed house. Someone might have loved the hilltop as much as we loved the valley farm. And I might have cherished my family at Mossy Dell, watching the kids ride their bikes around its shady courtyard from a porch overlooking that strangely compelling stage; one day Kathy would scatter my remains beneath the ashy-trunked oak on the bank above the pond. As I sell out, I can see this so clearly it seems real, as if in a parallel world it is happening. Another me, wiser—a man at peace with himself—had been graced with a loving wife, two fine kids, and the prettiest farm in Athens County, Ohio, and he'd had the sense to hunker down and love his life. There's an odd, fleeting refuge in imagining that outcome.

But Mossy Dell hadn't been our destination after all. I've lost my dream farm, and soon we'll sell our second-choice farm and move away, maybe closer to the kids; Claire is working in Chicago, and Tom is studying philosophy there. We haven't raised farmers, clearly, though we've given a couple of bright kids some things to think about. Parents seldom teach their children what they intend. If we're lucky, children take from us our best, the things we didn't say, the things we lived.

There are such mysteries in people. How did I understand my father as a man whose integrity went bone deep? Why did I inherit his burden? All I know is what he seemed to escape near the end, what reverberates in me to this day but which must end with me: a shotgun blast in a sleeping Michigan house.

In my mind's eye I see Dad walking into his nursery under the Florida sun. His right leg, withered from chronic spinal pain, he throws forward from the hip; his blue eyes are focused on the day's work; his scarred heart is barely beating. I see the trees in Mossy Dell's farmyard, and walking beneath their vault of boughs, a couple, no longer young but unaware of that, and their two children chasing a white terrier puppy.

In our last days on the hilltop, an image from my Georgia boyhood returns, a sacred moment. I'm four or five and run to a slight rise in our yard where butterflies blur the air. They engulf me as I turn and turn among their beating wings. They're different colors, improbably flying together, a dazzling airstream of orange and blue and yellow. Butterflies fill the air— endless butterflies—and infuse me with wonder and joy. Because I'm so young, I can't name, but only receive, their gift: a revelation of life's unfolding daily abundance: a miracle.

ACKNOWLEDGMENTS

WRITING A BOOK IS AN EXTENDED MEDITATION THAT, AS IN THIS case, can stretch into years. Like farming, it's an individual effort, but the help one receives is crucial, and I owe many people my sincere thanks. In Athens farming circles I'm especially indebted to Rick Duff and Sylvia Zimmerman, who helped me become a farmer, and to Don and Steve Shingler, who extended friendship to a newcomer. Friends from the larger sheep world, in particular Laura Fortmeyer, Charles Parker, and Jim Morgan, schooled me in livestock husbandry. A breeding-stock customer from West Virginia who appears in this book, John Stenger, took an interest in my project and shepherded me toward a balanced view of Appalachia.

I'm indebted to Patsy Sims, director of the Goucher College program in creative nonfiction, and to her writing mentors there who helped me turn notes and magazine articles into more personal prose: Richard Todd, Leslie Rubinkowski, Joe Mackall, Suzannah Lessard, and Diana Hume George. My friends and former colleagues at Ohio University Press, especially David Sanders, Gillian Berchowitz, Nancy Basmajian, and Jeff Kallet, took an interest in my extracurricular writing and suggested tweaks to what I shared. Later, at Otterbein University, teaching colleague Beth Daugherty, who is from Appalachian Ohio, helped me more fairly translate my angst in early chapters; and Candyce Canzoneri, a gifted wordsmith, read the manuscript at least twice, sharpening my prose and helping me see the humor. Then novelist and memoirist Bill Roorbach, of Maine, showed me how to meld linked essays into a memoir; his edit was a master class in narrative technique, and I flourished under his tutelage and belief in my story. Author Ana Maria Spagna played cleanup hitter, and dispensed tough love. At Michigan State University Press, I owe special thanks to Julie Loehr for her deft touch—and for saying yes.

Treasured friend David Bailey, a character in this book, also kindly read it, literally for years, serving—as he did in the events portrayed—as a sounding board. My siblings, Meg, David, and Peter, helped by reading, commenting, and answering my endless questions. So did my late mother, Rosie Gilbert, who sat for interviews about her life, early married years, and ranching adventures with my father. As I read a chapter to her as she lay dying, she observed that it dragged and needed cutting. She was right. Mom taught me how to plant trees, raise baby chicks, and tell stories. My father, Charles Churchill Gilbert, graced me with gifts as well: a love for grass farming, a respect for the ruminant, and a romantic heart.

Dad's children from his first marriage, Chuck Gilbert and Ann Wylie, shared memories of their early lives in California with Dad, their mother, and my mother. Ann's husband, John V. Wylie, a writer himself, was my most patient reader and enthusiastic cheerleader. I've been deeply influenced for thirty years by John's ideas about the evolution within humans of a force for goodness that can only be called God, and I'm humbled to have shared this vision I've made mine as well. You are a true friend, John, and I cannot possibly thank you enough.

Of course my longest-suffering supporters were my immediate family, my wife Kathy Krendl and our children, Claire and Tom, who endured years of drafts, discussions, and doubts—all to become characters in a book that wasn't their obsession, about an activity that had been mostly my obsession. I'm thankful to them and grateful for them, my partners in that adventure and on this journey.

AUTHOR'S NOTE

THIS IS A WORK OF NONFICTION. I HAVE RE-CREATED THE PAST BY consulting farm records and diaries, e-mails and letters, a stack of annotated old calendars, piles of receipts, photograph albums, my shaky memory, and my wife's better one. Aside from family members and a few friends, I have changed most names and some details to protect the privacy of people who didn't ask to enter my life, let alone to be written about. I have tried to be accurate and fair, and if I've made errors—trivial, I hope—I apologize to anyone affected.

Excerpts from this memoir have appeared, mostly in very different form, in literary and farming journals. I would like to thank for their editorial support *Brevity*; *Chautauqua*; *Farming: People, Land, Community*; *Fourth Genre*; *Memoir (and)*; *River Teeth*; *Sheep Canada*; *The Shepherd: A Guide to Sheep and Farm Life*; and *SNReview*.

ABOUT THE AUTHOR

RICHARD GILBERT AND HIS FAMILY LIVED FOR THIRTEEN YEARS IN the Appalachian foothills of southeastern Ohio and operated a sheep farm there for ten of those years. He and his wife, Kathy Krendl, now live in Westerville, Ohio, where they grow tomatoes, kayak on Alum Creek, and work for Otterbein University. They often spend weekends at a cabin in the Hocking Hills, just north of Athens and Mossy Dell Farm. Richard sometimes dreams about his old flock.